新文科·特色创新课程系列教材

本书受上海市高水平地方高校（学科）建设项目资助

Lecture Notes on
Computer Network Experiment
(Bilingual)

计算机网络与通信实训教程
（双语）

王学光　著

吉林大学出版社

·长　春·

图书在版编目(CIP)数据

计算机网络与通信实训教程 / 王学光著. —长春：
吉林大学出版社，2023.11
ISBN 978-7-5768-2737-8

Ⅰ. ①计… Ⅱ. ①王… Ⅲ. ①计算机网络－教材②计
算机通信－教材 Ⅳ. ①TP393②TN91

中国国家版本馆 CIP 数据核字(2023)第 239249 号

书　　名：计算机网络与通信实训教程(双语)
JISUANJI WANGLUO YU TONGXIN SHIXUN JIAOCHENG

作　　者：王学光
策划编辑：黄国彬
责任编辑：甄志忠
责任校对：赵　莹
装帧设计：姜　文
出版发行：吉林大学出版社
社　　址：长春市人民大街 4059 号
邮政编码：130021
发行电话：0431－89580028/29/21
网　　址：http://www.jlup.com.cn
电子邮箱：jldxcbs@sina.com
印　　刷：天津鑫恒彩印刷有限公司
开　　本：787mm×1092mm　　1/16
印　　张：13.75
字　　数：200 千字
版　　次：2024 年 5 月　第 1 版
印　　次：2024 年 5 月　第 1 次
书　　号：ISBN 978-7-5768-2737-8
定　　价：68.00 元

明德崇法　华章正铸

——华东政法大学"十四五"规划教材系列总序

教材不同于一般的书籍，它是传播知识的主要载体，体现着一个国家、一个民族的价值体系，是教师教学、学生学习的重要工具，更是教师立德树人的重要途径。一本优秀的教材，不仅是教师教学实践经验和学科研究成果的完美结合，更是教师展开思想教育和价值引领的重要平台。一本优秀的教材，也不只是给学生打下专业知识的厚实基础，更是通过自身的思想和语言的表达，引导学生全方位地成长。

习近平总书记深刻指出："当代中国的伟大社会变革，不是简单延续我国历史文化的母版，不是简单套用马克思主义经典作家设想的模板，不是其他国家社会主义实践的再版，也不是国外现代化发展的翻版。"新时代教材建设应当把体现党和国家的意志放在首位，要立足中华民族的价值观念，时刻把培养能够承担民族发展使命的时代新人作为高校教师编写教材的根本使命。为此，编写出一批能够体现中国立场、中国理论、中国实践、中国话语的有中国特色的高质量原创性教材，为培养德智体美劳全面发展的社会主义接班人和建设者提供保障，是高校教师的责任。

华东政法大学建校 70 年以来，一直十分注重教材的建设。特别是 1979 年第二次复校以来，与北京大学出版社、法律出版社、上海人民出版社等合作，先后推出了"高等学校法学系列教材""法学通用系列教材"法"学案例与图表系列教材""英语报刊选读系列教材""研究生教学系列用书""海商法系列教材""新世纪法学教材"等，其中曹建明教授主编的《国际经济法学概论》、苏惠渔教授主编的《刑法学》等教材荣获了司法部普通高校法学优秀教材一等奖；

史焕章研究员主编的《犯罪学概论》、丁伟教授主编的《冲突法论》、何勤华教授与魏琼教授编著的《西方商法史》及我本人主编的《诉讼证据法学》等教材荣获了司法部全国法学教材与科研成果二等奖;苏惠渔教授主编的《刑法学》、何勤华教授主编的《外国法制史》获得了上海市高校优秀教材一等奖;孙潮教授主编的《立法学》获得"九五"普通高等教育国家级重点教材立项;杜志淳教授主编的《司法鉴定实验教程》、何勤华教授主编的《西方法律思想史(第二版)》和《外国法制史(第五版)》、高富平教授与黄武双教授主编的《房地产法学(第二版)》、高富平教授主编的《物权法讲义》、余素青教授主编的《大学英语教程:读写译(1—4)》、苗伟明副教授主编的《警察技能实训教程》等分别入选第一批、第二批"十二五"普通高等教育本科国家级规划教材;王立民教授副主编的《中国法制史(第二版)》荣获首届全国优秀教材二等奖。1996年以来,我校教师主编的教材先后获得上海市级优秀教材一等奖、二等奖、三等奖共计72项。2021年,由何勤华教授主编的《外国法制史(第六版)》、王迁教授主编的《知识产权法教程(第六版)》、顾功耘教授主编的《经济法教程(第三版)》、王莲峰教授主编的《商标法学(第三版)》以及我本人主编的《刑事诉讼法学(第四版)》等5部教材获评首批上海高等教育精品教材,受到了广大师生的好评,取得了较好的社会效果和育人效果。

进入新时代,我校以习近平新时代中国特色社会主义思想铸魂育人为主线,在党中央"新工科、新医科、新农科、新文科"建设精神指引下,配合新时代背景下新法科、新文科建设的需求,根据学校"十四五"人才培养规划,制定了学校"十四五"教材建设规划。这次的教材规划一方面力求巩固学校优势学科专业,做好经典课程和核心课程教材建设的传承工作,另一方面适应新时代的人才培养需求和教育教学新形态的发展,推动教材建设的特色探索和创新发展,促进教学理念和内容的推陈出新,探索教学方式和方法的改革。

基于以上理念,围绕新文科建设,配合新法科人才培养体系改革和一流学科专业建设,在原有教材建设的基础上,我校展开系统化设计和规划,针对法学专业打造"新法科"教材共3个套系,针对非法学专业打造"新文科"教材共2个套系。"新法科"教材的3个套系分别是:"新法科·法学核心课程系列教材"新"法科·法律实务和案例教学系列教材""新法科·涉外法治人才培养系列教材"。"新文科"教材的 2 个套系分别是："新文科·经典传承系列教材"和"新文科·特色创新课程系列教材"。

"新法科"建设的目标，就是要解决传统法学教育存在的"顽疾"，培养与时代相适应的"人工智能＋法律"的复合型人才。这些也正是"新法科"3 套系列教材的设计初心和规划依据。

"新法科·法学核心课程系列教材"以推进传统的基础课程和核心课程的更新换代为目标，促进法学传统的基础和核心课程体系的改革。"新法科"理念下的核心课程教材系列，体现了新时代对法学传统的基础和核心课程建设的新要求，通过对我国司法实践中发生的大量新类型的法律案件的梳理、总结，开阔学生的法律思维，提升学生适用法律的能力。

"新法科·法律实务和案例教学系列教材"响应国家对于应用型、实践型人才的培养需要，以法律实务和案例教学的课程建设为基础，推进法学实践教学体系创新。此系列教材注重理论与实践的融合，旨在培养真正能够解决社会需求的应用型人才；以"新现象""新类型""新问题"为挑选案例的标准和基本原则，以培养学生学习兴趣、提升学生实践能力为导向。通过概念与案例的结合、法条与案例的结合，从具体案件到抽象理论，让学生明白如何在实践中解决疑难复杂问题，体会情、理与法的统一。

"新法科·涉外法治人才培养系列教材"针对培养具有国际视野和家国情怀、通晓国际规则、能够参与国际法律事务、善于维护国家利益、勇于推动全球治理体系变革的高素质涉外法治人才的培养目标，以涉外法治人才培养相关课程为基础，打造具有华政特色的涉外法治人才培养系列教材。

"新文科·经典传承系列教材"以政治学与行政学、公共事业管理、经济学、金融学、新闻学、汉语言文学、文化产业管理等专业的基础和主干课程为基础，在教材建设上，一方面体现学科专业特色，另一方面力求传统学科专业知识体系的现代创新和转型，注重把学科理论与新的社会文化问题、新的时代变局相联结，引导学生学习经典知识体系，以用于分析和思考新问题、解决新问题。

"新文科·特色创新课程系列教材"以各类创新、实践、融合等课程为基础，体现了"新文科"建设提出的融合创新、打破学科壁垒，实现跨学科、多学科交叉融合发展的理念,在教材建设上突破"小文科"思维,构建"大文科"格局,打造具有华政特色的各类特色课程系列教材。

华东政法大学 2022 年推出的这 5 个系列教材,在我看来,都有如下鲜明的特点:

第一,理论创新。系列教材改变了陈旧的理论范式,建构具有创新价值的知识体系,反映了学科专业理论研究最新成果,体现了经济社会和科技发展对人才培养提出的新要求。

第二,实践应用。系列教材的编写紧密围绕社会和文化建设中亟须解决的新问题,紧扣法治国家、法治政府、法治社会建设新需求,探索理论与实践的结合点,让教学实践服务于国家和社会的建设。

第三,中国特色。系列教材编写的案例和素材均来自于中国的法治建设和改革开放实践,传承并诠释了中国优秀传统文化,较好地体现了中国立场、中国理论、中国实践、中国话语。

第四,精品意识。为保证系列教材的高质量出版,我校遴选了各学科专业领域教学经验丰富、理论造诣深厚的学科带头人担任教材主编,选派优秀的中青年科研骨干参与教材的编写,组成教材编写团队,形成合力,为打造出高质量的精品教材提供保障。

当然,由于我校"新文科""新法科"的建设实践积累还不够丰厚,加之编写时间和编写水平有限,系列教材难免存在诸多不足之处。希望各位方家不吝赐教,我们将虚心听取,日后逐步完善。我希望,本系列教材的出版,可以为我国"新文科""新法科"建设贡献华政人的智慧。

是为序。

华东政法大学校长、教授　叶青

2022 年 8 月 22 日于华政园

目 录

Lecture noteson computer network experiment

1　实训环境介绍

1.1　实验室布局、拓扑结构图及相关说明

华东政法大学网络与信息安全实验室拓扑图（见图 1-1）。

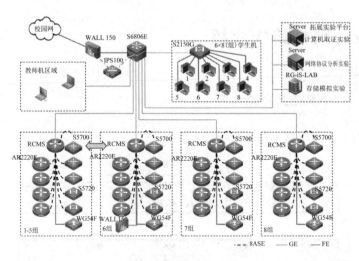

图 1-1　实验室拓扑图

每组机柜由 4 台 AR2220E 路由器，2 台 S5720-36C-EI-AC 交换机，3 台 S5700-28P-LI-AC 组成。具体机柜摆放图如图 1-2 所示。

前门

22	控制设备	22
21		21
20	配线架	20
19	理线架	19
18		18
17	AR2220E	17
16		16
15	AR2220E	15
14		14
13	AR2220E	13
12		12
11	AR2220E	11
10		10
9	S5720-36C-EI-AC	9
8		8
7	S5720-36C-EI-AC	7
6		6
5	S5700-28P-LI-AC	5
4		4
3	S5700-28P-LI-AC	3
2		2
1	S5700-28P-LI-AC	1

图 1-2　机柜摆放图

1.2　硬件结构介绍

1.2.1　AR2220E 路由器

AR2220E 路由器如图 1-3 所示。

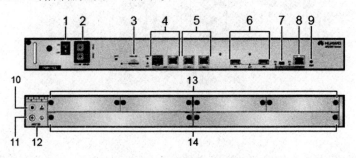

图 1-3　AR2220E 路由器

图注：

(1)电源开关

(2) 直流电源线接口

说明：

使用直流电源线缆将设备连接到外部电源。

(3) Micro SD 卡插槽

(4) WAN 侧接口：GE Combo 接口

(5) WAN 侧接口：2 个 GE 电接口

(6) 2 个 USB 接口（host）

说明：

插入 3G USB modem 时，建议安装 USB 塑料保护罩（选配）对它进行防护，USB 接口上方的 2 个螺钉孔用来固定 USB 塑料保护罩。USB 塑料保护罩的外观如下图所示。

(7) MiniUSB 接口

说明：

MiniUSB 接口和 Console 接口同时只能一个接口使能。

(8) CON/AUX 接口

说明：

AR2220-DC 不支持 AUX 功能。

(9) RST 按钮

注意：

复位按钮，用于手工复位设备。

复位设备会导致业务中断，需慎用复位按钮。

(10) ESD 插孔

说明：

对设备进行维护操作时，需要佩戴防静电腕带，防静电腕带的一端要插在 ESD 插孔里。

(11) 接地点

说明：

使用接地线缆将设备可靠接地，防雷、防干扰。

(12) 产品型号丝印

(13)4 个 SIC 槽位

(14)2 个 WSIC 槽位

1.2.2　S5700-28P-LI-AC 交换机

S5700-28P-LI-AC 交换机如图 1-4 所示。

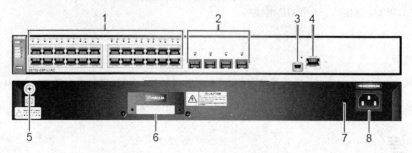

图 1-4　S5700-28P-LI-AC 交换机

图注:

(1)24 个 10/100/1000BASE-T 以太网电接口

(2)4 个 1000BASE-X 以太网光接口

支持的模块和线缆:

GE 光模块

GE-CWDM 彩色光模块

GE-DWDM 彩色光模块

GE 光电模块(V200R002C00 版本及以后版本支持,支持 10M/100M/1000M 速率)

堆叠光模块(V200R007C00 版本及以后版本支持)

1m、10m SFP+高速电缆

3m、10m AOC 光线缆(V200R003C00 版本及以后版本支持)

(3)1 个 MiniUSB 接口

(4)1 个 Console 接口

(5)接地螺钉

说明:

配套使用接地线缆。

(6)RPS 电源插座

说明:

配套使用 RPS 线缆，RPS 线缆不支持热插拔。

(7)交流端子防脱扣插孔

说明：

为安装交流端子防脱扣预留的插孔，交流端子防脱扣不随设备发货。

(8)交流电源插座

说明：

配套使用交流电源线缆

1.2.3　S5720-36C-EI-AC 交换机

S5720-36C-EI-AC 交换机如图 1-5 所示。

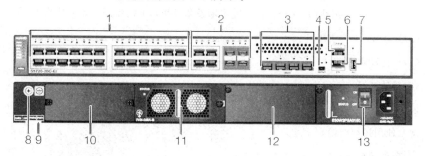

图 1-5　S5720-36C-EI-AC 交换机

图注：

(1)24 个 10/100/1000BASE-T 以太网电接口

(2)4 个 Combo 接口(10/100/1000BASE-T＋100/1000BASE-X)

Combo 光口支持的模块：

FE 光模块

GE 光模块

GE-CWDM 彩色光模块

GE-DWDM 彩色光模块

(3)4 个 10GE SFP＋以太网光接口

支持的模块和线缆：

GE 光模块

GE-CWDM 彩色光模块

GE-DWDM 彩色光模块

GE 光电模块(仅支持千兆速率)

10GE SFP＋光模块(不支持 OSXD22N00)

10GE-CWDM 光模块

10GE-DWDM 光模块(V200R009C00 版本及以后版本支持)

1m、3m、10m SFP＋高速电缆

5m SFP＋高速电缆(V200R009C00 版本及以后版本支持)

3m、10m AOC 光线缆

(4)1 个 MiniUSB 接口

(5)1 个 Console 接口

说明：

配套使用 Console 线缆。Console 线缆不随设备发货，如需使用，需单独购买。

(6)1 个 ETH 管理接口

(7)1 个 USB 接口

(8)接地螺钉

说明：

配套使用接地线缆。

(9)序列号标签

说明：

可抽出查看交换机的序列号和 MAC 地址信息。

(10)后插卡槽位

说明：

支持的插卡：

ES5D21X02S01(2 接口 10GE SFP＋光接口后插卡，S5720-EI 系列使用)

ES5D21X02T01(2 接口 10GBASE-T RJ45 电接口后插卡，S5720-EI 系列使用)

ES5D21VST000(2 接口 QSFP＋专用堆叠后插卡，S5720-EI 系列使用)

(11)风扇模块槽位

说明：

支持的风扇模块：FAN-028A-B 风扇模块

(12)电源模块槽位 2

说明：

支持的电源模块：

150W 交流电源模块

150W 直流电源模块

(13)电源模块槽位 1

说明：

支持的电源模块：

150W 交流电源模块

150W 直流电源模块

1.3　实验设备图谱说明

实验设备图谱如图 1-6 所示。

图 1-6　实验设备图谱

1.4　各试验台登录过程说明

1.4.1　背景信息

当完成 Console 用户界面的配置时，就可以通过 Console 口登录设备了。假如配置设备使用默认的 Console 用户界面属性，认证方式采用默认的 AAA 认证方式，登录过程如下所示。

1.4.2　操作步骤

(1)将 Console 通信电缆的 DB9(孔)插头插入 PC 机的串口(COM)中，再将 RJ-45 插头端插入设备的 Console 口中，如图 1-7 所示。

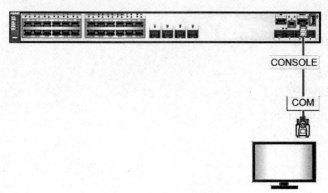

图 1-7　Console 口连接

（2）在 PC 上打开终端仿真软件，新建连接，设置连接的接口以及通信参数。（此处使用第三方软件 SecureCRT 为例进行介绍）

a. 如图 1-8 所示，单击" "，新建连接。

图 1-8　PC 终端仿真软件

b. 如图 1-9 所示，设置连接的接口以及通信参数。

连接的接口请根据实际情况进行选择。例如，在 Windows 系统中，可以通过在"设备管理器"中查看端口信息，选择连接的接口。

图 1-9　Quick Connect

设置终端软件的通信参数需与设备的缺省值保持一致，分别为：传输速率为 9600bit/s、8 位数据位、1 位停止位、无校验和无流控。

说明：

在缺省情况下，设备没有流控方式。RTS/CTS 缺省情况下处于使能状态，因此需要将该选项勾选掉，否则终端界面中无法输入命令行。

（3）单击"Connect"，直至系统出现如下显示，提示用户输入用户名和密码。（密码认证时，提示输入密码，以下显示信息仅为示意）

Login authentication
Username：admin
Password：
<HUAWEI>

（4）进入设备后，用户可以键入命令，对设备进行配置，需要帮助可以随时键入"？"。

1.4.3　检查配置结果

执行 display users[all]命令，查看用户界面的用户登录信息。

执行 display user-interface console 0 命令，查看用户界面信息。

执行 display local-user 命令，查看本地用户的属性信息。

执行 display access-user 命令，查看在线连接的用户信息。

2　设备基本配置

2.1　使用设备的命令行管理界面

设备提供丰富的功能，相应地也提供了多样的配置和查询命令。为便于用户使用这些命令，华为交换机按功能分类将命令分别注册在不同的命令行视图下。配置某一功能时，需首先进入对应的命令行视图，然后执行相应的命令进行配置。

设备提供的命令视图有很多，下面提到的视图是最常用的视图。其他视图的进入方式在具体的命令中都有说明，此处不再逐一介绍。

2.1.1　用户界面介绍

系统支持的用户界面有 Console 用户界面和 VTY 用户界面。

当用户通过 CLI 方式登录设备时，系统会分配一个用户界面用来管理、监控设备和用户间的当前会话。每个用户界面有对应的用户界面视图（user-interface view），在用户界面视图下网络管理员可以配置一系列参数，比如认证模式、用户级别等，当用户使用该用户界面登录时，将受到这些参数的约束，从而达到统一管理各种用户会话连接的目的。

设备支持两种类型的用户界面：

Console 用户界面：用来管理和监控通过 Console 口登录的用户。设备提供 Console 口，端口类型为 EIA/TIA-232 DCE。用户终端的串行口可以与设备 Console 口直接连接，实现对设备的本地访问。通过 MiniUSB 口登录设备使用的也是 Console 界面。

虚拟类型终端 VTY(virtual type terminal)用户界面：用来管理和监控通过 VTY 方式登录的用户。用户通过终端与设备建立 Telnet 或 STelnet 连接后，即建立了一条 VTY 通道。目前每台设备最多支持 15 个 VTY 用户同时访问。

2.1.2 用户与用户界面的关系

用户界面与用户并没有固定的对应关系。用户界面的管理和监控对象是使用某种方式登录的用户，虽然单个用户界面某一时刻只有一个用户使用，但它并不针对某个用户。

用户登录时，系统会根据用户的登录方式，自动给用户分配一个当前空闲的、编号最小的某类型的用户界面，整个登录过程将受该用户界面视图下配置的约束。比如用户 A 使用 Console 口登录设备时，将受到 Console 用户界面视图下配置的约束，当使用 VTY 1 登录设备时，将受到 VTY 1 用户界面视图下配置的约束。同一用户登录的方式不同，分配的用户界面不同；同一用户登录的时间不同，分配的用户界面可能不同。

📖 说明：

如果某 VTY 用户界面两次出现设备长时间不响应的情况，该 VTY 用户界面将被锁定，用户可以通过其他 VTY 用户界面登录，设备重启后可恢复。

2.1.3 用户界面的编号

用户界面的编号包括以下两种方式：

1. 相对编号

相对编号方式的形式是：用户界面类型＋编号。

此种编号方式只能唯一指定某种类型的用户界面中的一个或一组。相对编号方式遵守的规则如下：

Console 用户界面的编号：CON 0。

VTY 用户界面的编号：第一个为 VTY 0，第二个为 VTY 1，依此类推。

2. 绝对编号

使用绝对编号方式，可以唯一地指定一个用户界面或一组用户界面。使用命令 display user-interface 可查看到设备当前支持的用户界面以及它们的绝对编号。

对于一台设备，Console 口用户界面只有一个，但 VTY 类型的用户界面有 20 个，可以在系统视图下使用 user-interface maximum-vty 命令设置最大用户界面个数，其缺省值为 5。VTY 16～VTY 20 一直存在于系统中，不受 user-interface maximum-vty 命令的控制。

在缺省情况下，Console、VTY 用户界面在系统中的绝对编号，如表 2-1 所示。

表 2-1　用户界面的编号

	说明	绝对编号	相对编号
Console 用户界面	用来管理和监控通过 Console 口或 MiniUSB 口登录的用户	0	0
VTY 用户界面	用来管理和监控通过 Telnet 或 STelnet 方式登录的用户	34～48，50～54 其中 49 保留，50～54 为网管预留编号	第一个为 VTY 0，第二个为 VTY 1，依此类推。缺省存在 VTY 0～4 通道
			绝对编号 34～48 对应相对编号 VTY 0～VTY 14
			绝对编号 50～54 对应相对编号 VTY 16～VTY 20
			其中 VTY 15 保留，VTY 16～VTY 20 为网管预留编号
			只有当 VTY 0～VTY 14 全部被占用，且用户配置了 AAA 认证的情况下才可以使用 VTY 16～VTY 20

2.1.4　用户界面的用户认证

配置用户界面的用户认证方式后，用户登录设备时，系统对用户的身份进行认证。

对用户的认证有三种方式：AAA 认证、Password 认证和 None 认证。

AAA 认证：登录时需输入用户名和密码。设备根据配置的 AAA 用户名和密码验证用户输入的信息是否正确，如果正确，允许登录，否则拒绝登录。

Password 认证：也称密码认证，登录时需输入正确的认证密码。如果用户输入的密码与设备配置的认证密码相同，允许登录，否则拒绝登录。

None 认证：也称不认证，登录时不需要输入任何认证信息，可直接登录设备。

⚠ 注意：

为了保证更好的安全性，建议不要使用 None 认证方式。

无论何种验证方式，当用户登录设备失败时，系统会启动延时登录机制。首次登录失败后，延时 5 秒才可再次登录，后续登录失败次数每增加一次，延时时间增加 5 秒，即第 2 次登录失败延时 10 秒，第 3 次登录失败延时 15 秒。

2.1.5　用户界面的用户级别

系统支持对登录用户进行分级管理。用户所能访问命令的级别由用户的级别决定。

如果对用户采用 Password 认证或 None 认证，登录到设备的用户所能访问的命令级别由登录时的用户界面级别决定。

如果对用户采用 AAA 认证，登录到设备的用户所能访问的命令级别由 AAA 配置信息中本地用户的级别决定。

2.1.6 常用的命令行视图

常用命令行视图如表 2-2 所示。

表 2-2 常用命令行视图

视图名称	进入视图	视图功能
用户视图	用户从终端成功登录至设备即进入用户视图，在屏幕上显示 ＜HUAWEI＞	在用户视图下，用户可以完成查看运行状态和统计信息等功能
系统视图	在用户视图下，输入命令 system-view 后回车，进入系统视图 ＜HUAWEI＞ system-view Enter system view，return user view with Ctrl＋Z [HUAWEI]	在系统视图下，用户可以配置系统参数以及通过该视图进入其他的功能配置视图
接口视图	使用 interface 命令并指定接口类型及接口编号可以进入相应的接口视图 [HUAWEI] interface gigabitethernet X/Y/Z [HUAWEI-GigabitEthernetX/Y/Z] X/Y/Z 为需要配置的接口的编号，分别对应"堆叠 ID/子卡号/接口序号" 上述举例中 GigabitEthernet 接口仅为示意	配置接口参数的视图称为接口视图。在该视图下可以配置接口相关的物理属性、链路层特性及 IP 地址等重要参数

用户在使用命令行时，可以使用在线帮助以获取实时帮助，从而无须记忆大量的复杂的命令。

在线帮助通过键入"?"来获取，在命令行输入过程中，用户可以随时键入"?"以获得在线帮助。命令行在线帮助可分为完全帮助和部分帮助。

2.1.7　完全帮助

当用户输入命令时，可以使用命令行的完全帮助获取全部关键字和参数的提示。下面给出几种完全帮助的实例供参考：

在任一命令视图下，键入"?"获取该命令视图下所有的命令及其简单描述。举例如下：

```
<HUAWEI> ?
User view commands：
backup          Backup electronic elabel
cd              Change current directory
check           Check information
clear           Clear information
clock           Specify the system clock
compare         Compare function
…
```

键入一条命令的部分关键字，后接以空格分隔的"?"，如果该位置为关键字，则列出全部关键字及其简单描述。举例如下：

```
<HUAWEI> system-view
[HUAWEI] user-interface vty 0 4
[HUAWEI-ui-vty0-4] authentication-mode ?
aaa    AAA authentication，and this authentication mode is recommended
none   Login without checking
password   Authentication through the password of a user terminal interface
[HUAWEI-ui-vty0-4] authentication-mode aaa ?
<cr>
[HUAWEI-ui-vty0-4] authentication-mode aaa
```

其中，"aaa"和"password"是关键字，"AAA authentication"和"Authentication through the password of a user terminal interface"是对关键字的描述。

"<cr>"表示该位置没有关键字或参数，直接键入回车即可执行。

键入一条命令的部分关键字，后接以空格分隔的"?"，如果该位置为参数，则列出有关的参数名和参数描述。举例如下：

```
<HUAWEI> system-view
[HUAWEI] ftp timeout ?
INTEGER< 1-35791 > The value of FTP timeout，the default value is 30 minutes
[HUAWEI] ftp timeout 35 ?
<cr>


[HUAWEI] ftp timeout 35
```

其中，"INTEGER<1-35791>"是参数取值的说明，"The value of FTP timeout，the default value is 30 minutes"是对参数作用的简单描述。

2.1.8 部分帮助

当用户输入命令时，如果只记得此命令关键字的开头一个或几个字符，可以使用命令行的部分帮助获取以该字符串开头的所有关键字的提示。下面给出几种部分帮助的实例供参考：

键入一字符串，其后紧接"?"，列出以该字符串开头的所有关键字。举例如下：

```
<HUAWEI> d?
debugging                              delete
dir                                    display
<HUAWEI> d
```

键入一条命令，后接一字符串紧接"?"，列出命令以该字符串开头的所有关键字。举例如下：

```
<HUAWEI> display b?
bpdu                                    bridge
buffer
```

输入命令的某个关键字的前几个字母，按下[Tab]键，可以显示出完整的关键字，前提是这几个字母可以唯一标示出该关键字，否则，连续按下[Tab]键，可出现不同的关键字，用户可以从中选择所需要的关键字。

2.2 查看设备的系统和配置信息

2.2.1 查看设备的 CPU 占用率的统计信息

```
<HUAWEI> display cpu-usage
CPU Usage Stat. Cycle：60（Second）
CPU Usage：20% Max：99%
CPU Usage Stat. Time：2013-10-23  10：04：45
CPU utilization for five seconds：5%：one minute：5%：five minutes：5%
Max CPU Usage Stat. Time：2013-10-21 16：14：00.


TaskName  CPU   Runtime(CPU Tick High/Tick Low)   Task Explanation
VIDL      80%   0/e3a150c    0      DOPRA IDLE
OS        10%   0/ bfb044    0      Operation System
1AGAGT    6%    0/           0      1AGAGT
AAA       2%    0/           1d4a   AAA Authen Account Authorize
ACL       1%    0/           13362  ACL Access Control List
```

ADPT	1%	0/	0	ADPT Adapter
AGNT	0%	0/	0	AGNTSNMP agent task
AGT6	0%	0/	0	AGT6SNMP AGT6 task
ALM	0%	0/	0	ALM Alarm Management
ALS	0%	0/	527a3e	ALS Loss of Signal
AM	0%	0/	232cf	AM Address Management
APP	0%	0/	0	APP
ARP	0%	0/	36582	ARP
ASFI	0%	0/	0	ASFI
ASFM	0%	0/	0	ASFM
BATT	0%	0/	0	BATT Main Task
BFD	0%	0/	100f36	BFD Bidirection Forwarding Detect
BFDA	0%	0/	0	BFDA BFD Adapter

2.2.2 查看当前设备的内存占用率信息

<HUAWEI> display memory-usage

Memory utilization statistics at 2008-12-15 15：17：42＋08：00

System Total Memory Is：394152720 bytes

Total Memory Used Is：130975664 bytes

Memory Using Percentage Is：33%

2.2.3 查看设备的部件信息

```
<HUAWEI> display device
S5700-52P-LI-AC's Device status：
Slot Sub Type              Online      Power       Register       Status   Role

0-S5700-52P-LI             Present     PowerOn     Registered     Normal   Master
```

2.2.4 查看设备的版本信息

```
<HUAWEI> display version
Huawei Versatile Routing Platform Software
VRP（R）software，Version 5.160（S6720 V200R010C00SPC300）
Copyright（C）2000-2016 HUAWEI TECH CO.，LTD
HUAWEI S6720-54C-EI-48S-AC Routing Switch uptime is 0 week，0 day，5
hours，8 minutes


ES5D2S50Q002 1(Master)    : uptime is 0 week，0 day，5 hours，6 minutes
DDR      Memory Size      : 2048          M bytes
FLASH    Memory Size      : 446           M bytes
Pcb          Version  : VER. B
BootROM  Version  : 020a. 0001
BootLoad  Version  : 020a. 0001
CPLD      Version  : 0108
Software   Version  : VRP（R）Software，Version 5. 160（V200R010C00SPC300）
CARD1 information
Pcb          Version  : ES5D21Q04Q01 VER. A
CPLD      Version  : 0105
```

```
PWR2 information
Pcb          Version        : PWR VER. A
FAN1 information
Pcb          Version        : NA
```

2.2.5 查看设备当前生效的配置参数

```
display current-configuration
```

3 交换相关实验

3.1 虚拟局域网 VLAN 建立

3.1.1 组网需求

如图 3-1 所示，某企业的交换机连接有很多用户，且相同业务用户通过不同的设备接入企业网络。

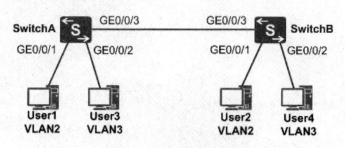

图 3-1 网络拓扑结构

为了通信的安全性，同时为了避免广播风暴，企业希望业务相同用户之间可以互相访问，业务不同用户之间不能直接访问。

可以在交换机上配置基于接口划分 VLAN，把业务相同的用户连接的接口划分到同一 VLAN。这样属于不同 VLAN 的用户不能直接进行二层通信，同一 VLAN 内的用户可以直接互相通信。

3.1.2　配置思路

采用如下的思路配置 VLAN：

(1)创建 VLAN 并将连接用户的接口加入 VLAN，实现不同业务用户之间的二层流量隔离。

(2)配置 SwitchA 和 SwitchB 之间的链路类型及通过的 VLAN，实现相同业务用户通过 SwitchA 和 SwitchB 通信。

3.1.3　操作步骤

(1)在 SwitchA 创建 VLAN2 和 VLAN3，并将连接用户的接口分别加入 VLAN。SwitchB 的配置与 SwitchA 类似，不再赘述。

```
<HUAWEI> system-view
[HUAWEI] sysname SwitchA
[SwitchA]vlan batch 2 3
[SwitchA]interface gigabitethernet 0/0/1
[SwitchA-GigabitEthernet0/0/1] port link-type access
[SwitchA-GigabitEthernet0/0/1] port default vlan 2
[SwitchA-GigabitEthernet0/0/1] quit
[SwitchA]interface gigabitethernet 0/0/2
[SwitchA-GigabitEthernet0/0/2] port link-type access
[SwitchA-GigabitEthernet0/0/2] port default vlan 3
[SwitchA-GigabitEthernet0/0/2] quit
```

(2)配置 SwitchA 与 SwitchB 连接的接口类型及通过的 VLAN。SwitchB 的配置与 SwitchA 类似，不再赘述。

```
[SwitchA]interface gigabitethernet 0/0/3
[SwitchA-GigabitEthernet0/0/3] port link-type trunk
[SwitchA-GigabitEthernet0/0/3] port trunk allow-pass vlan 2 3
```

（3）验证配置结果

将 User1 和 User2 配置在一个网段，比如 192.168.100.0/24；将 User3 和 User4 配置在一个网段，比如 192.168.200.0/24。

User1 和 User2 能够互相 Ping 通，但是均不能 Ping 通 User3 和 User4。User3 和 User4 能够互相 Ping 通，但是均不能 Ping 通 User1 和 User2。

3.1.4　配置文件

（1）SwitchA 的配置文件

```
#
sysname SwitchA
#
vlan batch 2 to 3
#
interface GigabitEthernet0/0/1
port link-type access
port default vlan 2
#
interface GigabitEthernet0/0/2
port link-type access
port default vlan 3
#
interface GigabitEthernet0/0/3
port link-type trunk
port trunk allow-pass vlan 2 to 3
#
return
```

(2)SwitchB 的配置文件

```
#
sysname SwitchB
#
vlan batch 2 to 3
#
interface GigabitEthernet0/0/1
port link-type access
port default vlan 2
#
interface GigabitEthernet0/0/2
port link-type access
port default vlan 3
#
interface GigabitEthernet0/0/3
port link-type trunk
port trunk allow-pass vlan 2 to 3
#
return
```

3.2　交换机端口隔离

3.2.1　组网需求

某企业研发办公室员工分为本公司员工、A 合作方公司员工和 B 合作方公司员工。如图 3-2 所示，PC1 和 PC2 分别代表 A、B 合作方员工，PC3 代表本公司研发员工，公司希望在节省 VLAN 资源的前提下，实现本公司员工和 A、B 两个合作方公司之间可以相互通信，但是 A、B 两个合作方公司员工之间无法通信。

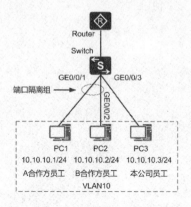

图 3-2　网络拓扑结构

3.2.2　配置思路

采用如下的思路配置端口隔离：

(1)配置接口加入 VLAN。

(2)设备缺省端口隔离为二层隔离三层互通，只需要将接口加入隔离组，就可以实现隔离组内接口之间二层数据的隔离。

3.2.3　操作步骤

1. 配置端口隔离功能

(1)配置 GE0/0/1 的端口隔离功能。

```
<HUAWEI> system-view
[HUAWEI] sysname Switch
[Switch]vlan 10
[Switch-vlan10]quit
[Switch]interface gigabitethernet 0/0/1
[Switch-GigabitEthernet0/0/1] port link-type access
```

```
[Switch-GigabitEthernet0/0/1] port default vlan 10
[Switch-GigabitEthernet0/0/1] port-isolate enable group 3
[Switch-GigabitEthernet0/0/1] quit
```

（2）配置 GE0/0/2 的端口隔离功能。

```
[Switch]interface gigabitethernet 0/0/2
[Switch-GigabitEthernet0/0/2] port link-type access
[Switch-GigabitEthernet0/0/2] port default vlan 10
[Switch-GigabitEthernet0/0/2] port-isolate enable group 3
[Switch-GigabitEthernet0/0/2] quit
```

（3）配置 GE0/0/3 加入 VLAN10。

```
[Switch]interface gigabitethernet 0/0/3
[Switch-GigabitEthernet0/0/3] port link-type access
[Switch-GigabitEthernet0/0/3] port default vlan 10
[Switch-GigabitEthernet0/0/3] quit
```

2. 验证配置结果

（1）PC1 和 PC2 数据报文不能互通。

（2）PC1 和 PC3 数据报文可以互通。

（3）PC2 和 PC3 数据报文可以互通。

3.2.4　配置文件

以下为 Switch 的配置文件。

```
#
sysname Switch
#
vlan batch 10
#
interface GigabitEthernet0/0/1
port link-type access
port default vlan 10
port-isolate enable group 3
#
interface GigabitEthernet0/0/2
port link-type access
port default vlan 10
port-isolate enable group 3
#
interface GigabitEthernet0/0/3
port link-type access
port default vlan 10
#
return
```

3.3　端聚合提供冗余备份链路

3.3.1　组网需求

如图 3-3 所示，SwitchA 和 SwitchB 通过以太链路分别连接 VLAN10 和 VLAN20 的网络，且 SwitchA 和 SwitchB 之间有较大的数据流量。

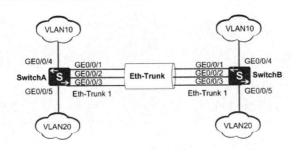

图 3-3　网络拓扑结构

用户希望 SwitchA 和 SwitchB 之间能够提供较大的链路带宽来使相同 VLAN 间互相通信。同时用户也希望能够提供一定的冗余度，保证数据传输和链路的可靠性。

3.3.2　配置思路

采用如下的思路配置负载分担链路聚合：

(1)创建 Eth-Trunk 接口并加入成员接口，实现增加链路带宽。

(2)创建 VLAN 并将接口加入 VLAN。

(3)配置负载分担方式，实现流量在 Eth-Trunk 各成员接口间的负载分担，增加可靠性。

3.3.3　操作步骤

(1)在 SwitchA 和 SwitchB 上创建 Eth-Trunk 接口并加入成员接口。

```
<HUAWEI> system-view
[HUAWEI] sysname SwitchA
[SwitchA] interface eth-trunk 1
[SwitchA-Eth-Trunk1] trunkport gigabitethernet 0/0/1 to 0/0/3
[SwitchA-Eth-Trunk1] quit
<HUAWEI> system-view
```

```
[HUAWEI] sysname SwitchB
[SwitchB] interface eth-trunk 1
[SwitchB-Eth-Trunk1] trunkport gigabitethernet 0/0/1 to 0/0/3
[SwitchB-Eth-Trunk1] quit
```

(2)创建 VLAN 并将接口加入 VLAN。

♯ 创建 VLAN10 和 VLAN20 并分别加入接口。SwitchB 的配置与 SwitchA 类似，不再赘述。

```
[SwitchA] vlan batch 10 20
[SwitchA] interface gigabitethernet 0/0/4
[SwitchA-GigabitEthernet0/0/4] port link-type trunk
[SwitchA-GigabitEthernet0/0/4] port trunk allow-pass vlan 10
[SwitchA-GigabitEthernet0/0/4] qui
[SwitchA] interface gigabitethernet 0/0/5
[SwitchA-GigabitEthernet0/0/5] port link-type trunk
[SwitchA-GigabitEthernet0/0/5] port trunk allow-pass vlan 20
[SwitchA-GigabitEthernet0/0/5] quit
```

♯ 配置 Eth-Trunk1 接口允许 VLAN10 和 VLAN20 通过。SwitchB 的配置与 SwitchA 类似，不再赘述。

```
[SwitchA] interface eth-trunk 1
[SwitchA-Eth-Trunk1] port link-type trunk
[SwitchA-Eth-Trunk1] port trunk allow-pass vlan 10 20
[SwitchA-Eth-Trunk1] quit
```

(3)配置 Eth-Trunk1 的负载分担方式。SwitchB 的配置与 SwitchA 类似，不再赘述。

```
[SwitchA] interface eth-trunk 1
[SwitchA-Eth-Trunk1] load-balance src-dst-mac
[SwitchA-Eth-Trunk1] quit
```

(4)验证配置结果。在任意视图下执行 display eth-trunk 1 命令，检查 Eth-Trunk 是否创建成功，成员接口是否正确加入。

```
[SwitchA] display eth-trunk 1
Eth-Trunk1's state information is：
Working Mode：NORMAL    Hash arithmetic：According to SA-XOR-DA
Least Active-linknumber：1    Max Bandwidth-affected-linknumber：8
Operate status：up    Number Of Up Port In Trunk：3
```

PortName	Status	Weight
GigabitEthernet0/0/1	Up	1
GigabitEthernet0/0/2	Up	1
GigabitEthernet0/0/3	Up	1

从以上信息看出 Eth-Trunk 1 中包含 3 个成员接口 GigabitEthernet0/0/1、GigabitEthernet0/0/2 和 GigabitEthernet0/0/3，成员接口的状态都为 Up。Eth-Trunk 1 的 Operate status 为 Up。

3.3.4 配置文件

1. SwitchA 的配置文件

```
#
sysname SwitchA
#
vlan batch 10 20
```

```
#
interface Eth-Trunk1
port link-type trunk
port trunk allow-pass vlan 10 20
load-balance src-dst-mac
#
interface GigabitEthernet0/0/1
eth-trunk 1
#
interface GigabitEthernet0/0/2
eth-trunk 1
#
interface GigabitEthernet0/0/3
eth-trunk 1
#
interface GigabitEthernet0/0/4
port link-type trunk
port trunk allow-pass vlan 10
#
interface GigabitEthernet0/0/5
port link-type trunk
port trunk allow-pass vlan 20
#
return
```

2. SwitchB 的配置文件

```
#
sysname SwitchB
#
```

```
vlan batch 10 20
#
interface Eth-Trunk1
port link-type trunk
port trunk allow-pass vlan 10 20
load-balance src-dst-mac
#
interface GigabitEthernet0/0/1
eth-trunk 1
#
interface GigabitEthernet0/0/2
eth-trunk 1
#
interface GigabitEthernet0/0/3
eth-trunk 1
#
interface GigabitEthernet0/0/4
port link-type trunk
port trunk allow-pass vlan 10
#
interface GigabitEthernet0/0/5
port link-type trunk
port trunk allow-pass vlan 20
#
return
```

3.4 生成树配置

3.4.1 组网需求

在一个复杂的网络中，网络规划者由于冗余备份的需要，一般都倾向于在设备之间部署多条物理链路，其中一条作主用链路，其他链路作备份。这样就难免会形成环形网络，若网络中存在环路，可能会引起广播风暴和 MAC 表项被破坏。

网络规划者规划好网络后，可以在网络中部署 STP 协议预防环路。当网络中存在环路，STP 通过阻塞某个端口以达到破除环路的目的。如图 3-4 所示，当前网络中存在环路，SwitchA、SwitchB、SwitchC 和 SwitchD 都运行 STP，通过彼此交互信息发现网络中的环路，并有选择地对某个端口进行阻塞，最终将环形网络结构修剪成无环路的树形网络结构，从而防止报文在环形网络中不断循环，避免设备由于重复接收相同的报文造成处理能力下降。

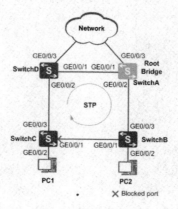

图 3-4　网络拓扑结构

3.4.2 配置思路

采用以下思路配置 STP 功能：

在处于环形网络中的交换设备上配置 STP 基本功能，包括：

(1)配置环网中的设备生成树协议工作在 STP 模式。

(2)配置根桥和备份根桥设备。

(3)配置端口的路径开销值，实现将该端口阻塞。

(4)使能 STP，实现破除环路。与 PC 相连的端口不用参与 STP 计算，将其设置为边缘端口并使能端口的 BPDU 报文过滤功能。

3.4.3 操作步骤

1. 配置 STP 基本功能

(1)配置环网中的设备生成树协议工作在 STP 模式。

♯ 配置交换设备 SwitchA 的 STP 工作模式。

```
<HUAWEI>system-view
[HUAWEI]sysname SwitchA
[SwitchA]stp mode stp
```

♯ 配置交换设备 SwitchB 的 STP 工作模式。

```
<HUAWEI>system-view
[HUAWEI]sysname SwitchB
[SwitchB]stp mode stp
```

♯ 配置交换设备 SwitchC 的 STP 工作模式。

```
<HUAWEI>system-view
[HUAWEI]sysname SwitchC
[SwitchC]stp mode stp
```

♯ 配置交换设备 SwitchD 的 STP 工作模式。

```
<HUAWEI>system-view
[HUAWEI]sysname SwitchD
[SwitchD]stp mode stp
```

（2）配置根桥和备份根桥设备

＃ 配置 SwitchA 为根桥。

```
[SwitchA]stp root primary
```

＃ 配置 SwitchD 为备份根桥。

```
[SwitchD]stp root secondary
```

（3）配置端口的路径开销值，实现将该端口阻塞。

＃ 端口路径开销值取值范围由路径开销计算方法决定，这里选择使用华为计算方法，配置将被阻塞端口的路径开销值为 20000。

＃ 同一网络内所有交换设备的端口路径开销应使用相同的计算方法。

· 配置 SwitchA 的端口路径开销计算方法为华为计算方法。

```
[SwitchA]stp pathcost-standard legacy
```

· 配置 SwitchB 的端口路径开销计算方法为华为计算方法。

```
[SwitchB]stp pathcost-standard legacy
```

· 配置 SwitchC 端口 GigabitEthernet0/0/1 端口路径开销值为 20000。

```
[SwitchC]stp pathcost-standard legacy
[SwitchC]interface gigabitethernet 0/0/1
[SwitchC-GigabitEthernet0/0/1]stp cost 20000
[SwitchC-GigabitEthernet0/0/1]quit
```

· 配置 SwitchD 的端口路径开销计算方法为华为计算方法。

```
[SwitchD]stp pathcost-standard legacy
```

（4）使能 STP，实现破除环路

＃ 将与 PC 机相连的端口设置为边缘端口并使能端口的 BPDU 报文过滤功能

• 配置 SwitchB 端口 GigabitEthernet0/0/2 为边缘端口并使能端口的 BPDU 报文过滤功能。

```
[SwitchB]interface gigabitethernet 0/0/2
[SwitchB-GigabitEthernet0/0/2]stp edged-port enable
[SwitchB-GigabitEthernet0/0/2]stp bpdu-filter enable
[SwitchB-GigabitEthernet0/0/2]quit
```

• 配置 SwitchC 端口 GigabitEthernet0/0/2 为边缘端口并使能端口的 BPDU 报文过滤功能。

```
[SwitchC]interface gigabitethernet 0/0/2
[SwitchC-GigabitEthernet0/0/2]stp edged-port enable
[SwitchC-GigabitEthernet0/0/2]stp bpdu-filter enable
[SwitchC-GigabitEthernet0/0/2]quit
```

♯ 设备全局使能 STP。

• 设备 SwitchA 全局使能 STP。

```
[SwitchA]stp enable
```

• 设备 SwitchB 全局使能 STP。

```
[SwitchB]stp enable
```

• 设备 SwitchC 全局使能 STP。

```
[SwitchC]stp enable
```

• 设备 SwitchD 全局使能 STP。

```
[SwitchD]stp enable
```

2. 验证配置结果

经过以上配置，在网络计算稳定后，执行以下操作，验证配置结果。

♯ 在 SwitchA 上执行 display stp brief 命令，查看端口状态和端口的保护类型，结果如下：

```
[SwitchA]display stp brief
MSTID   Port                      Role   STP State    Protection
  0     GigabitEthernet0/0/1      DESI   FORWARDING   NONE
  0     GigabitEthernet0/0/2      DESI   FORWARDING   NONE
```

将 SwitchA 配置为根桥后，与 SwitchB、SwitchD 相连的端口 GigabitEthernet0/0/2 和 GigabitEthernet0/0/1 在生成树计算中被选为指定端口。

♯ 在 SwitchB 上执行 display stp interface gigabitethernet 0/0/1 brief 命令，查看端口 GigabitEthernet0/0/1 状态，结果如下：

```
[SwitchB]display stp interface gigabitethernet 0/0/1 brief
MSTID   Port                      Role   STP State    Protection
  0     GigabitEthernet0/0/1      DESI   FORWARDING   NONE
```

端口 GigabitEthernet0/0/1 在生成树选举中成为指定端口，处于 FORWARDING 状态。

♯ 在 SwitchC 上执行 display stp brief 命令，查看端口状态，结果如下：

```
[SwitchC]display stp brief
MSTID   Port                      Role   STP State    Protection
  0     GigabitEthernet0/0/1      ALTE   DISCARDING   NONE
  0     GigabitEthernet0/0/3      ROOT   FORWARDING   NONE
```

端口 GigabitEthernet0/0/3 在生成树选举中成为根端口，处于 FORWARDING 状态。

端口 GigabitEthernet0/0/1 在生成树选举中成为 Alternate 端口，处于 DISCARDING 状态。

3.4.4　配置文件

1. SwitchA 的配置文件

```
#
sysname SwitchA
#
stp mode stp
stp instance 0 root primary
stp pathcost-standard legacy
#
return
```

2. SwitchB 的配置文件

```
#
sysname SwitchB
#
stp mode stp
stp pathcost-standard legacy
#
interface GigabitEthernet0/0/2
stp bpdu-filter enable
stp edged-port enable
#
return
```

3. SwitchC 的配置文件

```
#
sysname SwitchC
#
```

```
stp mode stp
stp pathcost-standard legacy
#
interface GigabitEthernet0/0/1
stp instance 0 cost 20000
#
interface GigabitEthernet0/0/2
stp bpdu-filter enable
stp edged-port enable
#
return
```

4. SwitchD 的配置文件

```
#
sysname SwitchD
#
stp mode stp
stp instance 0 root secondary
stp pathcost-standard legacy
#
return
```

4 路由相关实验

4.1 静态路由实验

4.1.1 组网需求

如图 4-1 所示，属于不同网段的主机通过几台 Router 相连，要求不配置动态路由协议，实现不同网段的任意两台主机之间能够互通。

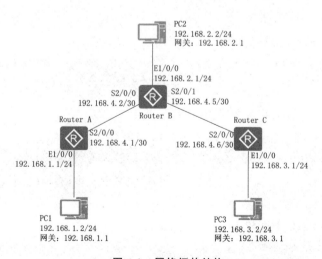

图 4-1 网络拓扑结构

4.1.2 配置思路

采用如下的思路配置 IPv4 静态路由：

(1)配置各路由器接口的 IP 地址，实现设备网络互通。

(2)在各主机上配置 IP 缺省网关，在各台路由器上配置 IP 静态路由及缺省路由，实现不配置动态路由协议，使不同网段的任意两台主机之间能够互通。

4.1.3 操作步骤

1. 配置各路由器接口的 IP 地址

♯ 在 RouterA 上配置接口 IP 地址，RouterB 和 RouterC 的配置与 RouterA 相同，此处省略。

```
[RouterA] interface Serial 2/0/0
[RouterA-Serial2/0/0] ip address 192.168.4.1 30
[RouterA-Serial2/0/0] quit
[RouterA] interface ethernet 1/0/0
[RouterA-Ethernet1/0/0] ip address 192.168.1.1 24
```

2. 配置静态路由

♯ 在 RouterA 上配置 IPv4 缺省路由。

```
[RouterA] ip route-static 0.0.0.0 0.0.0.0 192.168.4.2
```

♯ 在 RouterB 上配置两条 IPv4 静态路由。

```
[RouterB] ip route-static 192.168.1.0 255.255.255.0 192.168.4.1
[RouterB] ip route-static 192.168.3.0 255.255.255.0 192.168.4.6
```

♯ 在 RouterC 上配置 IPv4 缺省路由。

```
[RouterC] ip route-static 0.0.0.0 0.0.0.0 192.168.4.5
```

3. 配置主机

配置主机 PC1 的缺省网关为 192.168.1.1，主机 PC2 的缺省网关为 192.168.2.1，主机 PC3 的缺省网关为 192.168.3.1。

4. 验证配置结果

＃ 显示 RouterA 的 IP 路由表。

[RouterA] display ip routing-table
Route Flags：R - relay，D - download to fib

Routing Tables：Public
     Destinations：11   Routes：11

Destination/Mask	Proto	Pre	Cos	Flags	NextHop
Interface					
0.0.0.0/0	Static	60	0	RD	192.168.4.2
Serial2/0/0					
192.168.1.0/24	Direct	0	0	D	192.168.1.1
Ethernet1/0/0					
192.168.1.1/32	Direct	0	0	D	127.0.0.1
Ethernet1/0/0					
192.168.1.255/32	Direct	0	0	D	127.0.0.1
Ethernet1/0/0					
192.168.4.1/30	Direct	0	0	D	192.168.4.1
Serial2/0/0					
192.168.4.1/32	Direct	0	0	D	127.0.0.1
Serial2/0/0					
192.168.4.255/32	Direct	0	0	D	127.0.0.1
Serial2/0/0					
127.0.0.0/8	Direct	0	0	D	127.0.0.1
InLoopBack0					

127. 0. 0. 1/32	Direct	0	0	D	127. 0. 0. 1
InLoopBack0					
127. 255. 255. 255/32	Direct	0	0	D	127. 0. 0. 1
InLoopBack0					
255. 255. 255. 255/32	Direct	0	0	D	127. 0. 0. 1
InLoopBack0					

♯ 使用 Ping 命令验证连通性。

[RouterA] ping 192. 168. 3. 1

PING 192. 168. 3. 1: 56   data bytes，press CTRL _ C to break

Reply from 192. 168. 3. 1: bytes＝56 Sequence＝1 ttl＝254 time＝62 ms

Reply from 192. 168. 3. 1: bytes＝56 Sequence＝2 ttl＝254 time＝63 ms

Reply from 192. 168. 3. 1: bytes＝56 Sequence＝3 ttl＝254 time＝63 ms

Reply from 192. 168. 3. 1: bytes＝56 Sequence＝4 ttl＝254 time＝62 ms

Reply from 192. 168. 3. 1: bytes＝56 Sequence＝5 ttl＝254 time＝62 ms

--- 192. 168. 3. 1 ping statistics ---

5 packet(s) transmitted

5 packet(s) received

0. 00% packet loss

round-trip min/avg/max ＝ 62/62/63 ms

♯ 使用 Tracert 命令验证连通性。

[RouterA] tracert 192. 168. 3. 1

traceroute to   192. 168. 3. 1(192. 168. 3. 1)，max hops: 30 ，packet length:

40，press CTRL _ C to break

1 192. 168. 4. 2 31 ms   32 ms   31 ms

2 192. 168. 4. 6 62 ms   63 ms   62 ms

4.1.4 配置文件

1. RouterA 的配置文件

```
#
sysnameRouterA
#
interface Ethernet1/0/0
ip address 192.168.1.1 255.255.255.0
#
interface Serial2/0/0
ip address 192.168.4.1 255.255.255.252
#
ip route-static 0.0.0.0 0.0.0.0 192.168.4.2
#
return
```

2. RouterB 的配置文件

```
#
sysnameRouterB
#
interface Serial2/0/0
ip address 192.168.4.2 255.255.255.252
#
interface Serial2/0/1
ip address 192.168.4.5 255.255.255.252
#
interface E1/0/0
ip address 192.168.2.1 255.255.255.0
```

```
#
ip route-static 192.168.1.0 255.255.255.0 192.168.4.1
ip route-static 192.168.3.0 255.255.255.0 192.168.4.6
#
return
```

3. RouterC 的配置文件

```
#
sysnameRouterC
#
interface Ethernet1/0/0
ip address 192.168.3.1 255.255.255.0
#
interface Serial2/0/0
ip address 192.168.4.6 255.255.255.252
#
ip route-static 0.0.0.0 0.0.0.0 192.168.4.5
# return
```

4.2 RIP 路由协议实验

4.2.1 组网需求

如图 4-2 所示,在网络中有 4 台路由器,要求在 RouterA、RouterB、RouterC 和 RouterD 上实现网络互连。

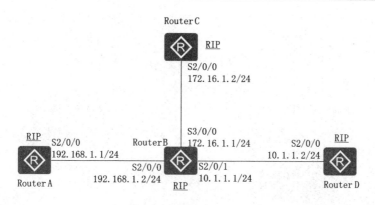

图 4-2 网络拓扑结构

4.2.2 配置思路

由于要在小型网络中实现设备的网络互连，所以推荐配置 RIP-2 路由协议。

(1)配置各接口 IP 地址，使网络可达。

(2)在各路由器上使能 RIP，基本实现网络互连。

(3)在各路由器上配置 RIP-2 版本，提升 RIP 路由扩展性能。

4.2.3 操作步骤

1. 配置各路由器接口的 IP 地址

配置 RouterA。

```
[RouterA] interface Serial 2/0/0
[RouterA-Serial2/0/0] ip address 192.168.1.1 24
```

RouterB、RouterC 和 RouterD 的配置与 RouterA 一致(略)。

2. 配置 RIP 基本功能

配置 RouterA。

```
[RouterA] rip
[RouterA-rip-1] network 192.168.1.0
[RouterA-rip-1] quit
```

配置 RouterB。

```
[RouterB] rip
[RouterB-rip-1] network 192.168.1.0
[RouterB-rip-1] network 172.16.0.0
[RouterB-rip-1] network 10.0.0.0
[RouterB-rip-1] quit
```

配置 RouterC。

```
[RouterC] rip
[RouterC-rip-1] network 172.16.0.0
[RouterC-rip-1] quit
```

配置 RouterD。

```
[RouterD] rip
[RouterD-rip-1] network 10.0.0.0
[RouterD-rip-1] quit
```

查看 RouterA 的 RIP 路由表。

```
[RouterA] display rip 1 route
Route Flags: R - RIP
             A - Aging, S - Suppressed, G - Garbage-collect

Peer 192.168.1.2   on Serial2/0/0
Destination/Mask        NextHop      Cost    Tag    Flags    Sec
```

| 10.0.0.0/8 | 192.168.1.2 | 1 | 0 | RA | 14 |
| 172.16.0.0/16 | 192.168.1.2 | 1 | 0 | RA | 14 |

从路由表中可以看出，RIP-1 发布的路由信息使用的是自然掩码。

3. 配置 RIP 的版本

在 RouterA 上配置 RIP-2。

```
[RouterA] rip
[RouterA-rip-1] version 2
[RouterA-rip-1] quit
```

在 RouterB 上配置 RIP-2。

```
[RouterB] rip
[RouterB-rip-1] version 2
[RouterB-rip-1] quit
```

在 RouterC 上配置 RIP-2。

```
[RouterC] rip
[RouterC-rip-1] version 2
[RouterC-rip-1] quit
```

在 RouterD 上配置 RIP-2。

```
[RouterD] rip
[RouterD-rip-1] version 2
[RouterD-rip-1] quit
```

4. 验证配置结果

查看 RouterA 的 RIP 路由表。

[RouterA] display rip 1 route

Route Flags：R - RIP

　　　　　　A - Aging，S - Suppressed，G - Garbage-collect

Peer 192. 168. 1. 2　onSerial2/0/0

Destination/Mask	NextHop	Cost	Tag	Flags	Sec
10. 1. 1. 0/24	192. 168. 1. 2	1	0	RA	32
172. 16. 1. 0/24	192. 168. 1. 2	1	0	RA	32

从路由表中可以看出，RIP-2 发布的路由中带有更为精确的子网掩码信息。

4. 2. 4　配置文件

1. RouterA 的配置文件

```
#
sysnameRouterA
#
interface Serial2/0/0
ip address 192. 168. 1. 1 255. 255. 255. 0
#
rip 1
version 2
network 192. 168. 1. 0
#
return
```

2. RouterB 的配置文件

```
#
sysnameRouterB
#
interface Serial2/0/0
ip address 192.168.1.2 255.255.255.0
#
interface Serial3/0/0
ip address 172.16.1.1 255.255.255.0
#
interface Serial2/0/1
ip address 10.1.1.1 255.255.255.0
#
rip 1
version 2
network 192.168.1.0
network 172.16.0.0
network 10.0.0.0
#
return
```

3. RouterC 的配置文件

```
#
sysnameRouterC
#
interface Serial2/0/0
ip address 172.16.1.2 255.255.255.0
#
```

```
rip 1
version 2
network 172.16.0.0
#
return
```

4. RouterD 的配置文件

```
#
sysnameRouterD
#
interfaceSerial2/0/1
ip address 10.1.1.2 255.255.255.0
#
rip 1
version 2
network 10.0.0.0
#
return
```

4.3 OSPF 路由协议实验

4.3.1 组网需求

如图 4-3 所示，所有的路由器都运行 OSPF，并将整个自治系统划分为 3 个区域，其中 RouterA 和 RouterB 作为 ABR 来转发区域之间的路由。配置完成后，每台路由器都应学到 AS 内的到所有网段的路由。

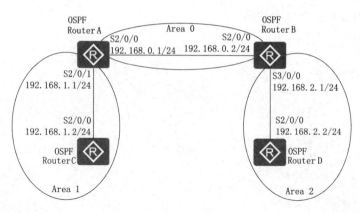

图 4-3　网络拓扑结构

4.3.2　配置思路

采用如下的思路配置 OSPF 基本功能：

(1)在各路由器上使能 OSPF。

(2)指定不同区域内的网段。

4.3.3　操作步骤

1. 配置各路由器接口的 IP 地址

♯ 配置 RouterA。

```
<Huawei>system-view
[Huawei]sysname RouterA
[RouterA] interface Serial 2/0/0
[RouterA-Serial2/0/0] ip address 192.168.0.1 24
[RouterA-Serial2/0/0] quit
[RouterA] interface Serial 2/0/1
[RouterA-Serial2/0/1] ip address 192.168.1.1 24
[RouterA-Serial2/0/1] quit
```

♯ RouterB、RouterC、RouterD 的配置方法与 RouterA 一致(略)

2. 配置 OSPF 基本功能

＃ 配置 RouterA。

```
[RouterA] router id 1. 1. 1. 1
[RouterA] ospf
[RouterA-ospf-1] area 0
[RouterA-ospf-1-area-0. 0. 0. 0] network 192. 168. 0. 0 0. 0. 0. 255
[RouterA-ospf-1-area-0. 0. 0. 0] quit
[RouterA-ospf-1] area 1
[RouterA-ospf-1-area-0. 0. 0. 1] network 192. 168. 1. 0 0. 0. 0. 255
[RouterA-ospf-1-area-0. 0. 0. 1] quit
[RouterA-ospf-1] quit
```

＃ 配置 RouterB。

```
[RouterB] router id 2. 2. 2. 2
[RouterB] ospf
[RouterB-ospf-1] area 0
[RouterB-ospf-1-area-0. 0. 0. 0] network 192. 168. 0. 0 0. 0. 0. 255
[RouterB-ospf-1-area-0. 0. 0. 0] quit
[RouterB-ospf-1] area 2
[RouterB-ospf-1-area-0. 0. 0. 2] network 192. 168. 2. 0 0. 0. 0. 255
[RouterB-ospf-1-area-0. 0. 0. 2] quit
[RouterB-ospf-1] quit
```

＃ 配置 RouterC。

```
[RouterC] router id 3. 3. 3. 3
[RouterC] ospf
[RouterC-ospf-1] area 1
[RouterC-ospf-1-area-0. 0. 0. 1] network 192. 168. 1. 0 0. 0. 0. 255
[RouterC-ospf-1-area-0. 0. 0. 1] quit
[RouterC-ospf-1] quit
```

＃ 配置 RouterD。

```
[RouterD] router id 4.4.4.4

[RouterD] ospf

[RouterD-ospf-1] area 2

[RouterD-ospf-1-area-0.0.0.2] network 192.168.2.0 0.0.0.255

[RouterD-ospf-1-area-0.0.0.2] quit

[RouterD-ospf-1] quit
```

3. 验证配置结果

＃ 查看 RouterA 的 OSPF 邻居。

```
[RouterA] display ospf peer

                    OSPF Process 1 with Router ID 1.1.1.1

                              Neighbors

Area 0.0.0.0 interface 192.168.0.1(Serial2/0/0)'s neighbors

Router ID: 2.2.2.2          Address: 192.168.0.2

State: Full   Mode: Nbr is  Master  Priority: 1

DR: 192.168.0.2   BDR: 192.168.0.1    MTU: 0

Dead timer due in 36   sec

Retrans timer interval: 5

Neighbor is up for 00: 15: 04

Authentication Sequence: [ 0 ]

                              Neighbors

Area 0.0.0.1 interface 192.168.1.1(Serial2/0/1)'s neighbors

Router ID: 3.3.3.3          Address: 192.168.1.2

State: Full   Mode: Nbr is  Master  Priority: 1

DR: 192.168.1.2   BDR: 192.168.1.1    MTU: 0

Dead timer due in 39   sec

Retrans timer interval: 5
```

Neighbor is up for 00：07：32

Authentication Sequence：［ 0 ］

♯ 显示 RouterA 的 OSPF 路由信息。

[RouterA] display ospf routing

OSPF Process 1 with Router ID 1.1.1.1

Routing Tables

Routing for Network

Destination	Cost	Type	NextHop	AdvRouter	Area
192.168.0.0/24	1	Transit	192.168.0.1	1.1.1.1	0.0.0.0
192.168.1.0/24	1	Transit	192.168.1.1	1.1.1.1	0.0.0.1
192.168.2.0/24	2	Inter-area	192.168.0.2	2.2.2.2	0.0.0.0

Total Nets：3

Intra Area：2 Inter Area：1 ASE：0 NSSA：0

♯ 显示 RouterA 的 LSDB。

[RouterA] display ospf lsdb

OSPF Process 1 with Router ID 1.1.1.1

Link State Database

Area：0.0.0.0

Type	LinkState ID	AdvRouter	Age	Len	Sequence	Metric
Router	2.2.2.2	2.2.2.2	317	48	80000003	1
Router	1.1.1.1	1.1.1.1	316	48	80000002	1
Network	192.168.0.2	2.2.2.2	399	32	800000F8	0
Sum-Net	192.168.2.0	2.2.2.2	237	28	80000002	1
Sum-Net	192.168.1.0	1.1.1.1	295	28	80000002	1

Area：0. 0. 0. 1

Type	LinkState ID	AdvRouter	Age	Len	Sequence	Metric
Router	3. 3. 3. 3	3. 3. 3. 3	217	60	80000008	1
Router	1. 1. 1. 1	1. 1. 1. 1	289	48	80000002	1
Network	192. 168. 1. 1	1. 1. 1. 1	202	28	80000002	0
Sum-Net	192. 168. 2. 0	1. 1. 1. 1	242	28	80000001	2
Sum-Net	192. 168. 0. 0	1. 1. 1. 1	300	28	80000001	1

♯ 查看 RouterD 的路由表，并使用 Ping 进行测试连通性。

〔RouterD〕display ospf routing

OSPF Process 1 with Router ID 4. 4. 4. 4

Routing Tables

Routing for Network

Destination	Cost	Type	NextHop	AdvRouter	Area
192. 168. 0. 0/24	2	Inter-area	192. 168. 2. 1	2. 2. 2. 2	0. 0. 0. 2
192. 168. 1. 0/24	3	Inter-area	192. 168. 2. 1	2. 2. 2. 2	0. 0. 0. 2
192. 168. 2. 0/24	1	Transit	192. 168. 2. 2	4. 4. 4. 4	0. 0. 0. 2

Total Nets：5

Intra Area：1　Inter Area：2　ASE：0　NSSA：0

〔RouterD〕ping 192. 168. 1. 2

PING 172. 16. 1. 1：56　data bytes，press CTRL _ C to break

Reply from192. 168. 1. 2：bytes＝56 Sequence＝1 ttl＝253 time＝62 ms

Reply from192. 168. 1. 2：bytes＝56 Sequence＝2 ttl＝253 time＝16 ms

Reply from192. 168. 1. 2：bytes＝56 Sequence＝3 ttl＝253 time＝62 ms

Reply from192. 168. 1. 2：bytes＝56 Sequence＝4 ttl＝253 time＝94 ms

Reply from192. 168. 1. 2：bytes＝56 Sequence＝5 ttl＝253 time＝63 ms

——192. 168. 1. 2 ping statistics ——

5 packet(s) transmitted

5 packet(s) received

0.00% packet loss

round-trip min/avg/max = 16/59/94 ms

4.3.4 配置文件

1. RouterA 的配置文件

```
#
sysnameRouterA
#
router id 1.1.1.1
#
interface Serial2/0/0
ip address 192.168.0.1 255.255.255.0
#
interface Serial2/0/1
ip address 192.168.1.1 255.255.255.0
#
ospf 1
area 0.0.0.0
network 192.168.0.0 0.0.0.255
area 0.0.0.1
network 192.168.1.0 0.0.0.255
#
return
```

2. RouterB 的配置文件

```
#
sysnameRouterB
#
router id 2. 2. 2. 2
#
interface Serial2/0/0
ip address 192. 168. 0. 2 255. 255. 255. 0
#
interface Serial2/0/1
ip address 192. 168. 2. 1 255. 255. 255. 0
#
ospf 1
area 0. 0. 0. 0
network 192. 168. 0. 0 0. 0. 0. 255
area 0. 0. 0. 2
network 192. 168. 2. 0 0. 0. 0. 255
#
return
```

3. RouterC 的配置文件

```
#
sysnameRouterC
#
router id 3. 3. 3. 3
#
interface Serial2/0/1
ip address 192. 168. 1. 2 255. 255. 255. 0
```

```
#
ospf 1
area 0. 0. 0. 1
network 192. 168. 1. 0 0. 0. 0. 255
#
return
```

4. RouterD 的配置文件

```
#
sysname RouterD
#
router id 4. 4. 4. 4
#
interface Serial2/0/1
ip address 192. 168. 2. 2 255. 255. 255. 0
#
ospf 1
area 0. 0. 0. 2
network 192. 168. 2. 0 0. 0. 0. 255
#
return
```

4.4　利用三层交换机实现相同和不同 VLAN 间的通信及隔离

4.4.1　组网需求

如图 4-4 所示，为了通信的安全性，某公司将访客、员工、服务器分别划分到 VLAN10、VLAN20、VLAN30 中。公司希望：员工、服务器主机、访客均能访问 Internet。访客只能访问 Internet，不能与其他任何 VLAN 的用户通信。员工 A 可以访问服务器区的所有资源，但其他员工只能访问服务器 A 的 21 端口(FTP 服务)。

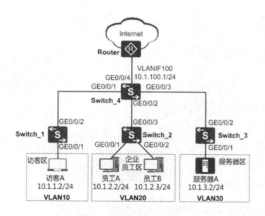

图 4-4　网络拓扑结构

4.4.2　配置思路

可采用如下思路配置通过流策略实现 VLAN 间互访控制：

(1)配置 VLAN 并将各接口加入 VLAN，使员工、服务器、访客间二层隔离。

(2)配置 VLANIF 接口及其 IP 地址，使员工、服务器、访客间可三层互通。

（3）配置上行路由，使员工、服务器、访客均可通过 Switch 访问 Internet。

（4）配置并应用流策略，使员工 A 可以访问服务器区的所有资源，其他员工只能访问服务器 A 的 21 端口，且只允许员工访问服务器；使访客只能访问 Internet。

4.4.3 操作步骤

1. 配置 VLAN 并将各接口加入 VLAN，使员工、服务器、访客间二层隔离

♯ 在 Switch _ 1 上创建 VLAN10，并将接口 GE0/0/1 以 Untagged 方式加入 VLAN10，接口 GE0/0/2 以 Tagged 方式加入 VLAN10。Switch _ 2 和 Switch _ 3 的配置与 Switch _ 1 类似，不再赘述。

```
<HUAWEI> system-view
[HUAWEI] sysname Switch _ 1
[Switch _ 1] vlan batch 10
[Switch _ 1] interface gigabitethernet 0/0/1
[Switch _ 1-GigabitEthernet0/0/1] port link-type access
[Switch _ 1-GigabitEthernet0/0/1] port default vlan 10
[Switch _ 1-GigabitEthernet0/0/1] quit
[Switch _ 1] interface gigabitethernet 0/0/2
[Switch _ 1-GigabitEthernet0/0/2] port link-type trunk
[Switch _ 1-GigabitEthernet0/0/2] port trunk allow-pass vlan 10
[Switch _ 1-GigabitEthernet0/0/2] quit
```

♯ 在 Switch _ 4 上创建 VLAN10、VLAN20、VLAN30、VLAN100，并配置接口 GE0/0/1~GE0/0/4 分别以 Tagged 方式加入 VLAN10、VLAN20、VLAN30、VLAN100。

```
<HUAWEI> system-view
[HUAWEI] sysname Switch _ 4
[Switch _ 4] vlan batch 10 20 30 100
[Switch _ 4] interface gigabitethernet 0/0/1
[Switch _ 4-GigabitEthernet0/0/1] port link-type trunk
[Switch _ 4-GigabitEthernet0/0/1] port trunk allow-pass vlan 10
[Switch _ 4-GigabitEthernet0/0/1] quit
[Switch _ 4] interface gigabitethernet 0/0/2
[Switch _ 4-GigabitEthernet0/0/2] port link-type trunk
[Switch _ 4-GigabitEthernet0/0/2] port trunk allow-pass vlan 20
[Switch _ 4-GigabitEthernet0/0/2] quit
[Switch _ 4] interface gigabitethernet 0/0/3
[Switch _ 4-GigabitEthernet0/0/3] port link-type trunk
[Switch _ 4-GigabitEthernet0/0/3] port trunk allow-pass vlan 30
[Switch _ 4-GigabitEthernet0/0/3] quit
[Switch _ 4] interface gigabitethernet 0/0/4
[Switch _ 4-GigabitEthernet0/0/4] port link-type trunk
[Switch _ 4-GigabitEthernet0/0/4] port trunk allow-pass vlan 100
[Switch _ 4-GigabitEthernet0/0/4] quit
```

2. 配置 VLANIF 接口及其 IP 地址，使员工、服务器、访客间可以三层互通

♯ 在 Switch _ 4 上创建 VLANIF10、VLANIF20、VLANIF30、VLANIF100，并分别配置其 IP 地址为 10.1.1.1/24、10.1.2.1/24、10.1.3.1/24、10.1.100.1/24。

```
[Switch _ 4] interface vlanif 10
[Switch _ 4-Vlanif10] ip address 10. 1. 1. 1 24
[Switch _ 4-Vlanif10] quit
[Switch _ 4] interface vlanif 20
[Switch _ 4-Vlanif20] ip address 10. 1. 2. 1 24
[Switch _ 4-Vlanif20] quit
[Switch _ 4] interface vlanif 30
[Switch _ 4-Vlanif30] ip address 10. 1. 3. 1 24
[Switch _ 4-Vlanif30] quit
[Switch _ 4] interface vlanif 100
[Switch _ 4-Vlanif100] ip address 10. 1. 100. 1 24
[Switch _ 4-Vlanif100] quit
```

3. 配置上行路由，使员工、服务器、访客均可通过 Switch 访问 Internet

在 Switch _ 4 上配置 OSPF 基本功能，发布用户网段以及 Switch _ 4 与 Router 之间的互联网段。

```
[Switch _ 4] ospf
[Switch _ 4-ospf-1] area 0
[Switch _ 4-ospf-1-area-0. 0. 0. 0] network 10. 1. 1. 0 0. 0. 0. 255
[Switch _ 4-ospf-1-area-0. 0. 0. 0] network 10. 1. 2. 0 0. 0. 0. 255
[Switch _ 4-ospf-1-area-0. 0. 0. 0] network 10. 1. 3. 0 0. 0. 0. 255
[Switch _ 4-ospf-1-area-0. 0. 0. 0] network 10. 1. 100. 0 0. 0. 0. 255
[Switch _ 4-ospf-1-area-0. 0. 0. 0] quit
[Switch _ 4-ospf-1] quit
```

📖 说明：

Router 上需要进行如下配置：

·将连接 Switch 的接口以 Tagged 方式加入 VLAN100，并指定 VLANIF100 的 IP 地址与 10.1.100.1 在同一网段。

·配置 OSPF 基本功能，并发布 Switch 与 Router 之间的互联网段。

4. 配置并应用流策略，控制员工、访客、服务器之间的访问

(1)通过 ACL 定义每个流。

♯ 在 Switch _ 4 上配置 ACL 3000，禁止访客访问员工区和服务器区。

```
[Switch _ 4] acl 3000
[Switch _ 4-acl-adv-3000] rule deny ip destination 10.1.2.1 0.0.0.255
[Switch _ 4-acl-adv-3000] rule deny ip destination 10.1.3.1 0.0.0.255
[Switch _ 4-acl-adv-3000] quit
```

♯ 在 Switch _ 4 上配置 ACL 3001，使员工 A 可以访问服务器区的所有资源，其他员工只能访问服务器 A 的 21 端口。

```
[Switch _ 4] acl 3001
[Switch _ 4-acl-adv-3001] rule permit tcp destination 10.1.3.2 0 destination-port eq 21
[Switch _ 4-acl-adv-3001] rule permit ip source 10.1.2.2 0 destination 10.1.3.1 0.0.0.255
[Switch _ 4-acl-adv-3001] rule deny ip destination 10.1.3.1 0.0.0.255
[Switch _ 4-acl-adv-3001] quit
```

(2)配置流分类，区分不同的流。

♯ 在 Switch _ 4 上创建流分类 c _ custom、c ˋ_ staff，并分别配置匹配规则 3000、3001。

```
[Switch _ 4] traffic classifier c _ custom
[Switch _ 4-classifier-c _ custom] if-match acl 3000
[Switch _ 4-classifier-c _ custom] quit
[Switch _ 4] traffic classifier c _ staff
[Switch _ 4-classifier-c _ staff] if-match acl 3001
[Switch _ 4-classifier-c _ staff] quit
```

（3）配置流行为，指定流动作。

＃ 在 Switch _ 4 上创建流行为 b1，并配置允许动作。

```
[Switch _ 4] traffic behavior b1
[Switch _ 4-behavior-b1] permit
[Switch _ 4-behavior-b1] quit
```

（4）配置流策略，关联流分类和流行为。

＃ 在 Switch _ 4 上创建流策略 p _ custom、p _ staff，并分别将流分类 c _ custom、c _ staff 与流行为 b1 关联。

```
[Switch _ 4] traffic policy p _ custom
[Switch _ 4-trafficpolicy-p _ custom] classifier c _ custom behavior b1
[Switch _ 4-trafficpolicy-p _ custom] quit
[Switch _ 4] traffic policy p _ staff
[Switch _ 4-trafficpolicy-p _ staff] classifier c _ staff behavior b1
[Switch _ 4-trafficpolicy-p _ staff] quit
```

（5）应用流策略，实现员工、访客、服务器之间的访问控制。

＃ 在 Switch _ 4 上，分别在 VLAN10、VLAN20 的入方向应用流策略 p _ custom、p _ staff。

```
[Switch _ 4] vlan 10
[Switch _ 4-vlan10] traffic-policy p _ custom inbound
[Switch _ 4-vlan10] quit
[Switch _ 4] vlan 20
[Switch _ 4-vlan20] traffic-policy p _ staff inbound
[Switch _ 4-vlan20] quit
```

5. 验证配置结果

配置访客 A 的 IP 地址为 10.1.1.2/24，缺省网关为 VLANIF10 接口的 IP 地址 10.1.1.1；配置员工 A 的 IP 地址为 10.1.2.2/24，缺省网关为 VLANIF20 接口的 IP 地址 10.1.2.1；配置员工 B 的 IP 地址为 10.1.2.3/24，缺省网关为 VLANIF20 接口的 IP 地址 10.1.2.1；配置服务器 A 的 IP 地址为 10.1.3.2/24，缺省网关为 VLANIF30 接口的 IP 地址 10.1.3.1。

配置完成后：

·访客 A 不能 Ping 通员工 A、服务器 A；员工 A 和服务器 A 不能 Ping 通访客 A。

·员工 A 可以 Ping 通服务器 A，即可以使用服务器 A 的 FTP 服务，也可以使用服务器 A 的。

·员工 B 可以 Ping 不通服务器 A，只能使用服务器 A 的 FTP 服务。

·访客、员工 A、员工 B、服务器 A 均可以 Ping 通 Router 连接 Switch _ 4 的接口的 IP 地址 10.1.100.2/24，也就都可以访问 Internet。

4.4.4　配置文件

1. Switch _ 1 的配置文件

```
#
sysname Switch _ 1
#
vlan batch 10
```

```
#
interface GigabitEthernet0/0/1
port link-type access
port default vlan 10
#
interface GigabitEthernet0/0/2
port link-type trunk
port trunk allow-pass vlan 10
#
return
```

2. Switch _ 2 的配置文件

```
#
sysname Switch _ 2
#
vlan batch 20
#
interface GigabitEthernet0/0/1
port link-type access
port default vlan 20
#
interface GigabitEthernet0/0/2
port link-type access
port default vlan 20
#
interface GigabitEthernet0/0/3
port link-type trunk
port trunk allow-pass vlan 20
#
return
```

3. Switch _ 3 的配置文件

```
#
sysname Switch _ 3
#
vlan batch 30
#
interface GigabitEthernet0/0/1
port link-type access
port default vlan 30
#
interface GigabitEthernet0/0/2
port link-type trunk
port trunk allow-pass vlan 30
#
return
```

4. Switch _ 4 的配置文件

```
#
sysname Switch _ 4
#
vlan batch 10 20 30 100
#
acl number 3000
rule 5 deny ip destination 10. 1. 2. 0 0. 0. 0. 255
rule 10 deny ip destination 10. 1. 3. 0 0. 0. 0. 255
acl number 3001
rule 5 permit tcp destination 10. 1. 3. 2 0 destination-port eq ftp
rule 10 permit ip source 10. 1. 2. 2 0 destination 10. 1. 3. 0 0. 0. 0. 255
rule 15 deny ip destination 10. 1. 3. 0 0. 0. 0. 255
```

```
#
traffic classifier c _ custom operator and
if-match acl 3000
traffic classifier c _ staff operator and
if-match acl 3001
#
traffic behavior b1
permit
#
traffic policy p _ custom match-order config
classifier c _ custom behavior b1
traffic policy p _ staff match-order config
classifier c _ staff behavior b1
#
vlan 10
traffic-policy p _ custom inbound
vlan 20
traffic-policy p _ staff inbound
#
interface Vlanif10
ip address 10. 1. 1. 1 255. 255. 255. 0
#
interface Vlanif20
ip address 10. 1. 2. 1 255. 255. 255. 0
#
interface Vlanif30
ip address 10. 1. 3. 1 255. 255. 255. 0
#
interface Vlanif100
```

```
ip address 10. 1. 100. 1 255. 255. 255. 0
#
interface GigabitEthernet0/0/1
port link-type trunk
port trunk allow-pass vlan 10
#
interface GigabitEthernet0/0/2
port link-type trunk
port trunk allow-pass vlan 20
#
interface GigabitEthernet0/0/3
port link-type trunk
port trunk allow-pass vlan 30
#
interface GigabitEthernet0/0/4
port link-type trunk
port trunk allow-pass vlan 100
#
ospf 1
area 0. 0. 0. 0
network 10. 1. 1. 0 0. 0. 0. 255
network 10. 1. 2. 0 0. 0. 0. 255
network 10. 1. 3. 0 0. 0. 0. 255
network 10. 1. 100. 0 0. 0. 0. 255
#
return
```

5 网络安全＋综合实验

5.1 利用动态 NAPT 实现局域网访问互联网

5.1.1 组网需求

如图 5-1 所示，某公司 A 区和 B 区的私网用户和互联网相连，路由器上接口 Serial 2/0/0 的公网地址为 202.169.10.1/24，对端运营商侧地址为 202.169.10.2/24。A 区用户希望使用公网地址池中的地址（202.169.10.100 ～202.169.10.200）采用 NAT 方式替换 A 区内部的主机地址（网段为 192.168.20.0/24），访问因特网。B 区用户希望结合 B 区的公网 IP 地址比较少的情况，使用公网地址池（202.169.10.80～202.169.10.83）采用 IP 地址和端口的替换方式替换 B 区内部的主机地址（网段为 10.0.0.0/24），访问因特网。

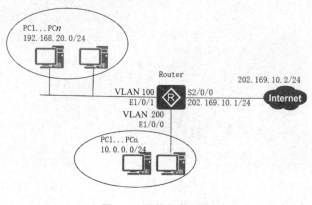

图 5-1　网络拓扑结构

5.1.2　配置思路

配置动态地址转换的思路为：配置接口 IP 地址、缺省路由和在 WAN 侧接口下配置 NAT Outbound，实现内部主机访问外网服务功能。

5.1.3　操作步骤

1. 在 Router 上配置接口 IP 地址

```
<Huawei> system-view
[Huawei] sysname Router
[Router]vlan 100
[Router-vlan100]quit
[Router]interface vlanif 100
[Router-Vlanif100]ip address 192.168.20.1 24
[Router-Vlanif100]quit
[Router]interface ethernet 1/0/1
[Router-Ethernet1/0/1] port link-type access
[Router-Ethernet1/0/1] port default vlan 100
[Router-Ethernet1/0/1] quit
[Router]vlan 200
[Router-vlan200]quit
[Router]interface vlanif 200
[Router-Vlanif200]ip address 10.0.0.1 24
[Router-Vlanif200]quit
[Router]interface ethernet 1/0/0
[Router-Ethernet1/0/0] port link-type access
[Router-Ethernet1/0/0] port default vlan 200
[Router-Ethernet1/0/0] quit
```

［Router］interface Serial 2/0/0

［Router-Serial2/0/0］ip address 202.169.10.1 24

［Router-Serial2/0/0］quit

    2. 在 Router 上配置缺省路由，指定下一跳地址为 202.169.10.2

［Router］ip route-static 0.0.0.0 0.0.0.0 202.169.10.2

    3. 在 Router 上配置 NAT Outbound

［Router］nat address-group 1 202.169.10.100 202.169.10.200

［Router］nat address-group 2 202.169.10.80 202.169.10.83

［Router］acl 2000

［Router-acl-basic-2000］rule 5 permit source 192.168.20.0 0.0.0.255

［Router-acl-basic-2000］quit

［Router］acl 2001

［Router-acl-basic-2001］rule 5 permit source 10.0.0.0 0.0.0.255

［Router-acl-basic-2001］quit

［Router］interface Serial 2/0/0

［Router-Serial2/0/0］nat outbound 2000 address-group 1 no-pat

［Router-Serial2/0/0］nat outbound 2001 address-group 2

［Router-Serial2/0/0］quit

📖 说明：

    如果需要在 Router 上执行 ping-a source-ip-address 命令通过指定发送 ICMP ECHO-REQUEST 报文的源 IP 地址来验证内网用户是否可以访问因特网，需要配置命令 ip soft-forward enhance enable 激活设备控制报文的增强转发功能，这样，私网的源地址才能通过 NAT 转换为公网地址。在缺省情况下，设备产生的控制报文的增强转发功能处于使能状态。如果之前已经执行命令 undo ip soft-forward enhance enable 去使能增强转发功能，需要重新在系统视图下执行命令 ip soft-forward enhance enable。

4. 验证配置结果

＃ 在 Router 上执行命令 display nat outbound，查看地址转换结果。

```
<Router>display nat outbound
NAT Outbound Information：
```

Interface	Acl	Address-group/IP/Interface	Type
Serial2/0/0	2000	1	no-pat
Serial2/0/0	2001	2	pat

```
Total：2
```

＃ 在 Router 上执行命令 ping，验证内网可以访问因特网。

```
<Router>ping -a 192.168.20.1 202.169.10.2
PING 202.169.10.2：56 data bytes, press CTRL _ C to break
Reply from 202.169.10.2：bytes＝56 Sequence＝1 ttl＝255 time＝1 ms
Reply from 202.169.10.2：bytes＝56 Sequence＝2 ttl＝255 time＝1 ms
Reply from 202.169.10.2：bytes＝56 Sequence＝3 ttl＝255 time＝1 ms
Reply from 202.169.10.2：bytes＝56 Sequence＝4 ttl＝255 time＝1 ms
Reply from 202.169.10.2：bytes＝56 Sequence＝5 ttl＝255 time＝1 ms
——202.169.10.2 ping statistics——
5 packet(s) transmitted
5 packet(s) received
0.00% packet loss
round-trip min/avg/max = 1/1/2 ms
<Router>ping -a 10.0.0.1 202.169.10.2
PING 202.169.10.2：56 data bytes, press CTRL _ C to break
Reply from 202.169.10.2：bytes＝56 Sequence＝1 ttl＝255 time＝1 ms
```

Reply from 202. 169. 10. 2：bytes＝56 Sequence＝2 ttl＝255 time＝1 ms

Reply from 202. 169. 10. 2：bytes＝56 Sequence＝3 ttl＝255 time＝1 ms

Reply from 202. 169. 10. 2：bytes＝56 Sequence＝4 ttl＝255 time＝1 ms

Reply from 202. 169. 10. 2：bytes＝56 Sequence＝5 ttl＝255 time＝1 ms

——202. 169. 10. 2 ping statistics——

5 packet(s) transmitted

5 packet(s) received

0. 00% packet loss

round-trip min/avg/max ＝ 1/1/2 ms

5.1.4 配置文件

Router 的配置文件

```
#
sysname Router
#
vlan batch 100 200
#
acl number 2000
rule 5 permit source 192.168.20.0 0.0.0.255
#
acl number 2001
rule 5 permit source 10.0.0.0 0.0.0.255
#
nat address-group 1 202.169.10.100 202.169.10.200
nat address-group 2 202.169.10.80 202.169.10.83
#
interface Vlanif100
```

```
ip address 192.168.20.1 255.255.255.0
#
interface Vlanif200
ip address 10.0.0.1 255.255.255.0
#
interface Ethernet1/0/1
port link-type access
port default vlan 100
#
interface Ethernet1/0/0
port link-type access
port default vlan 200
#
interface Serial2/0/0
ip address 202.169.10.1 255.255.255.0
nat outbound 2000 address-group 1 no-pat
nat outbound 2001 address-group 2
#
ip route-static 0.0.0.0 0.0.0.0 202.169.10.2
#
return
```

5.2   利用 NAT 实现外网访问内网服务器

5.2.1   组网需求

如图 5-2 所示，某公司的网络中提供 WWW Server 和 FTP Server 供外部网络用户访问。其中 WWW Server 的内部 IP 地址为 192.168.20.2/24，提供服务的端口为 8080，对外公布的地址为 202.169.10.5/24。FTP Server 的内部 IP 地址为 10.0.0.3/24，对外公布的地址为 202.169.10.33/24，对端运营商侧地址为 202.169.10.2/24。要求通过路由器的 NAT 功能把该公司的内部网络连接到因特网上。

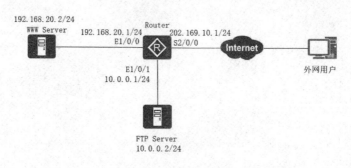

图 5-2   网络拓扑结构

5.2.2 配置思路

采用如下思路配置内部服务器：

(1)配置接口 IP 地址，并在接口 Serial 2/0/0 上配置 NAT Server，实现外部网络用户访问内网服务器功能。

(2)配置 Router 的缺省路由。

(3)使能 FTP 的 NAT ALG 功能，实现外部用户的 FTP 访问能正常穿越 NAT。

5.2.3 操作步骤

1. 在 Router 上配置接口 IP 地址和 NAT Server

```
<Huawei> system-view
[Huawei] sysname Router
[Router]vlan 100
```

```
[Router-vlan100]quit

[Router]interface vlanif 100

[Router-Vlanif100]ip address 192.168.20.1 24

[Router-Vlanif100]quit

[Router]interface ethernet 1/0/0

[Router-Ethernet1/0/0] port link-type access

[Router-Ethernet1/0/0] port default vlan 100

[Router-Ethernet1/0/0] quit

[Router]vlan 200

[Router-vlan200]quit

[Router]interface vlanif 200

[Router-Vlanif200]ip address 10.0.0.1 24

[Router-Vlanif200]quit

[Router]interface ethernet 1/0/1

[Router-Ethernet1/0/1] port link-type access

[Router-Ethernet1/0/1] port default vlan 200

[Router-Ethernet1/0/1] quit

[Router]interface Serial 2/0/0

[Router-Serial2/0/0] ip address 202.169.10.1 24

[Router-Serial2/0/0] nat server protocol tcp global 202.169.10.5 www inside
192.168.20.2 8080

[Router-Serial2/0/0] nat server protocol tcp global 202.169.10.33 ftp inside
10.0.0.2 ftp

[Router-Serial2/0/0] quit
```

2. 在 Router 上配置缺省路由，下一跳地址为 202.169.10.2

```
[Router]ip route-static 0.0.0.0 0.0.0.0 202.169.10.2
```

3. 在 Router 上使能 FTP 的 NAT ALG 功能

[Router]nat alg ftp enable

4. 验证配置结果

♯ 在 Router 上执行 display nat server 操作，结果如下。

<Router>display nat server

Nat Server Information：

Interface：Serial 2/0/0

Global IP/Port   : 202.169.10.5/80(www)

Inside IP/Port   : 192.168.20.2/8080

Protocol ：6(tcp)

VPN instance-name   : ————————

Acl number   : ————————

Vrrp id   : ————————

Description   : ————————

Global IP/Port   : 202.169.10.33/21(ftp)

Inside IP/Port   : 10.0.0.2/21(ftp)

Protocol ：6(tcp)

VPN instance-name   : ————————

Acl number   : ————————

Vrrp id   : ————————

Description   : ————————

Total ：2

♯ 在 Router 上执行 display NAT alg 操作，结果如下。

```
<Router>display nat alg
NAT Application Level Gateway Information：


Application              Status


dns                      Disabled
ftp                      Enabled
rtsp                     Disabled
sip                      Disabled
pptp                     Disabled
```

＃ 验证外网用户是否能正常访问公司的 WWW Server 和 FTP Server
（略）。

5.2.4 配置文件

Router 的配置文件

```
#
sysname Router
#
vlan batch 100 200
#
nat alg ftp enable
#
interface Vlanif100
ip address 192. 168. 20. 1 255. 255. 255. 0
```

```
#
interface Vlanif200
ip address 10. 0. 0. 1 255. 255. 255. 0
#
interface Ethernet1/0/0
port link-type access
port default vlan 100
#
interface Ethernet1/0/1
port link-type access
port default vlan 200
#
interfaceSerial 2/0/0
ip address 202. 169. 10. 1 255. 255. 255. 0
nat server protocol tcp global 202. 169. 10. 5 www inside 192. 168. 20. 2 8080
nat server protocol tcp global 202. 169. 10. 33 ftp inside 10. 0. 0. 2 ftp
#
ip route-static 0. 0. 0. 0 0. 0. 0. 0 202. 169. 10. 2
#
return
```

5.3 IP 访问列表实验

5.3.1 组网需求

如图 5-3 所示，某个对外提供 Web、FTP 和 Telnet 服务的企业通过 Router 的接口 Serial 2/0/0 访问外部网络，通过 Router 的接口 Eth1/0/0 加入

VLAN。已知企业的网段为 202.169.10.0/24，企业内部的 WWW 服务器、FTP 服务器 和 Telnet 服务器 IP 地 址 分 别 为 202.169.10.5/24、202.169.10.6/24 和 202.169.10.7/24。

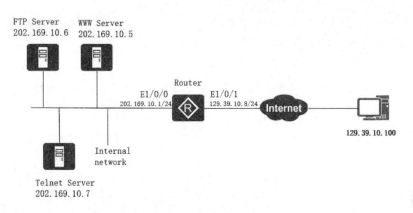

图 5-3　网络拓扑结构

为了实现内部网络具备较高的安全性，企业希望在 Router 上配置防火墙功能，使外部网络只有特定用户可以访问内部服务器，企业内只有内部服务器可以访问外部网络。

5.3.2　配置思路

采用如下的配置思路：

(1)为企业内部网络和外部网络配置不同的安全区域。

(2)配置安全域间，在安全域间使用防火墙。

(3)配置不同的高级 ACL，对可以访问内部服务器的外部网络用户以及可以访问外部网络的内部服务器进行分类。

(4)在安全域间配置基于高级 ACL 的包过滤。

5.3.3　操作步骤

1. 配置安全区域

♯ 为企业内部网络配置安全区域。

```
<Huawei> system-view
[Huawei] sysname Router
[Router] firewall zone company
[Router-zone-company] priority 12
[Router-zone-company] quit
```

＃配置接口加入 VLAN，并配置 VLANIF 接口的 IP 地址，将接口
VLANIF 100 加入安全区域 company。

```
[Router] vlan batch 100
[Router] interface ethernet 1/0/0
[Router-Ethernet1/0/0] port link-type access
[Router-Ethernet1/0/0] port default vlan 100
[Router-Ethernet1/0/0] quit
[Router] interface vlanif 100
[Router-Vlanif100] ip address 202.169.10.1 255.255.255.0
[Router-Vlanif100] zone company
[Router-Vlanif100] quit
```

＃为外部网络配置安全区域。

```
[Router] firewall zone external
[Router-zone-external] priority 5
[Router-zone-external] quit
```

＃将接口 Serial 2/0/0 加入安全区域 external。

```
[Router] interface ethernet 1/0/1
[Router-Serial2/0/0] ip address 129.39.10.8 255.255.255.0
[Router-Serial2/0/0] zone external
[Router-Serial2/0/0] quit
```

2. 配置安全域间

```
[Router] firewall interzone company external
[Router-interzone-company-external] firewall enable
[Router-interzone-company-external] quit
```

3. 配置 ACL 3001

＃ 创建 ACL 3001。

```
[Router] acl 3001
```

＃ 配置允许特定用户从外部网络可以访问内部服务器。

```
[Router-acl-adv-3001] rule permit tcp source 202.39.2.3 0.0.0.0 destination
202.169.10.5 0.0.0.0
[Router-acl-adv-3001] rule permit tcp source 202.39.2.3 0.0.0.0 destination
202.169.10.6 0.0.0.0
[Router-acl-adv-3001] rule permit tcp source 202.39.2.3 0.0.0.0 destination
202.169.10.7 0.0.0.0
```

＃ 配置其他用户不能从外部网络访问企业内部的任何主机。

```
[Router-acl-adv-3001] rule deny ip
[Router-acl-adv-3001] quit
```

4. 配置 ACL 3002

＃ 创建 ACL 3002。

```
[Router] acl 3002
```

＃ 配置允许内部服务器访问外部网络。

```
[Router-acl-adv-3002] rule permit ip source 202.169.10.5 0.0.0.0
[Router-acl-adv-3002] rule permit ip source 202.169.10.6 0.0.0.0
[Router-acl-adv-3002] rule permit ip source 202.169.10.7 0.0.0.0
```

♯ 配置网络内部的其他用户不能访问外部网络。

```
[Router-acl-adv-3002] rule deny ip
[Router-acl-adv-3002] quit
```

5. 在安全域间配置基于高级 ACL 的包过滤

```
[Router] firewall interzone company external
[Router interzone-company-external] packet-filter 3001 inbound
[Router-interzone-company-external] packet-filter 3002 outbound
[Router-interzone-company-external] quit
```

6. 验证配置结果

♯ 配置成功后，仅特定主机(129.39.10.100)可以访问内部服务器，仅内部服务器可以访问外部网络。

♯ 在 Router 上执行 display firewall interzone [zone-name1 zone-name2] 操作，结果如下。

```
[Router] display firewall interzone company external
interzone company external
firewall enable
packet-filter default deny inbound
packet-filter default permit outbound
packet-filter 3001 inbound
packet-filter 3002 outbound
```

5.3.4 配置文件

Router 的配置文件

```
♯
sysname Router
```

```
#
vlan batch 100
#
acl number 3001
rule 5 permit tcp source 202. 39. 2. 3 0 destination 202. 169. 10. 5 0
rule 10 permit tcp source 202. 39. 2. 3 0 destination 202. 169. 10. 6 0
rule 15 permit tcp source 202. 39. 2. 3 0 destination 202. 169. 10. 7 0
rule 20 deny ip
acl number 3002
rule 5 permit ip source 202. 169. 10. 5 0
rule 10 permit ip source 202. 169. 10. 6 0
rule 15 permit ip source 202. 169. 10. 7 0
rule 20 deny ip
#
interface Vlanif100
ip address 202. 169. 10. 1 255. 255. 255. 0
zone company
#
firewall zone company
priority 12
#
firewall zone external
priority 5
#
firewall interzone company external
firewall enable
packet-filter 3001 inbound
packet-filter 3002 outbound
```

```
#
interface Ethernet1/0/0
port link-type access
port default vlan 100
#
interface Ethernet1/0/1
ip address 129.39.10.8 255.255.255.0
zone external
#
return
```

6　华为交换机和路由器内部命令、使用格式和服务器搭建介绍

6.1　通用

表 6-1　通用命令

CISCO	HUAWEI	描述
no	undo	取消、关闭当前设置
show	display	显示查看
exit	quit	退回上级
hostname	sysname	设置主机名
en，config terminal	system-view	进入全局模式
delete	delete	删除文件
reload	reboot	重启
write	save	保存当前配置
username	local-user	创建用户
shutdown	shutdown	禁止、关闭端口
show version	display version	显示当前系统版本
show startup-config	display saved-configuration	查看已保存过的配置
show running-config	display current-configuration	显示当前配置
no debug all	Ctrl＋D	取消所有 DEBUG 命令
erase startup-config	reset saved-configuration	删除配置
end	return	退到用户视图
exit	logout	登出
logging	info-center	指定信息中心配置信息

续表

CISCO	HUAWEI	描述
line	user-interface	进入线路配置(用户接口)模式
start-config	saved-configuration	启动配置
running-config	current-configuration	当前配置
host	ip host	host 名字和 ip 地址对应

6.2 交换

表 6-2 交换命令

CISCO	HUAWEI	描述
enable password	set authentication password simple	配置明文密码
interface type/number	interface type/number	进入接口
interface vlan 1	interface vlan 1	进入 VLAN 配置 VLAN 管理地址
interface rang	interface ethID to ID	定议多个端口的组
enable secret	super password	设置特权口令
duplex (half | full | auto)	duplex (half | full | auto)	配置接口状态
speed (10/100/1000)	speed (10/100/1000)	配置端口速率
switchport mode trunk	port link-type trunk	配置 Trunk
vlan ID /no vlan ID	vlan batch ID /undo vlan batch ID	添加、删除 VLAN
switchport access vlan	port access vlan ID	将端口接入 VLAN
show interface	display interface	查看接口
show vlan ID	display vlan ID	查看 VLAN
encapsulation	link-protocol	封装协议
channel-group 1 mode on	port link-aggregation group 1	链路聚合
ip routing	default open	开启三层交换的路由功能

续表

CISCO	HUAWEI	描述
no switchport	nen-support	开启接口三层功能
vtp domain	GVRP	对跨以太网交换机的 VLAN 进行动态注册和删除
spanning-tree vlan ID root primary	stp instance id root primary	STP 配置根网桥
spanning-tree vlan ID priority	stp primary vlaue	配置网桥优先级
show spanning-tree	dis stp brief	查看 STP 配置

6.3 路由

表 6-3 路由命令

CISCO	HUAWEI	描述
ip route 0.0.0.0 0.0.0.0	ip route-static 0.0.0.0 0.0.0.0	配置默认路由
ip route 目标网段＋掩码 下一跳	ip route-static 目标网段＋掩码 下一跳	配置静态路由
show ip route	display ip routing-table	查看路由表
router rip /network 网段	rip /network 网段	启用 RIP，并宣告网段
router ospf	ospf	启用 OSPF
network ip 反码 area <area-id>	area <area-id>	配置 OSPF 区域
no auto-summary	rip split-horizon	配置 RIP V2 水平分割
show ip protocol	display ip protocol	查看路由协议
access-list 1-99 permit/deny ip	rule id permit source ip	标准访问控制列表

CISCO	HUAWEI	描述
access-list 100-199 permit/deny protocol source ip ＋ 反码 destination ip＋反码 operator operan	rule ｛normal｜special｝ ｛permit｜deny｝｛tcp｜udp｝ source ｛＜ip wild＞｜any｝ destination ＜ip wild＞｜ any｝［operate］	扩展访问控制列表
ip nat inside source static	nat server global ＜ip＞ ［port］inside ＜ip＞ port ［protocol］	配置静态地址转换

6.4 Web 服务器搭建方法

本书实验涉及的 Web 服务器中采用 EasyServer 软件进行搭建，具体方法如下：

(1)双击运行软件程序 🖳 MyWebServer.exe ，出现如图 6-1 所示界面。

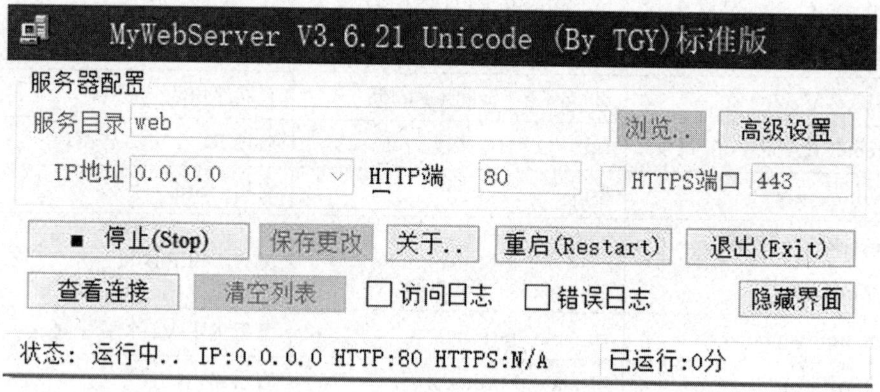

图 6-1

(2)点击"停止(Stop)"，如图 6-2 所示；

图 6-2

(3)根据实验中服务器的端口号，修改"HTTP 端口"参数，如实验中为
8080 端口，则修改如图 6-3 所示。

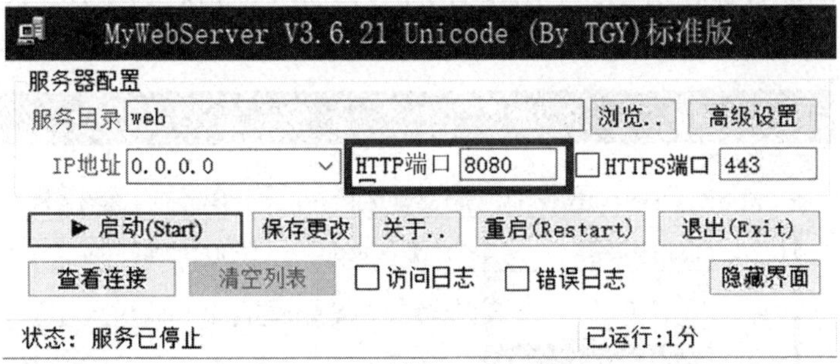

图 6-3

(4)修改服务目录，点击"浏览"选项，如图 6-4 所示。

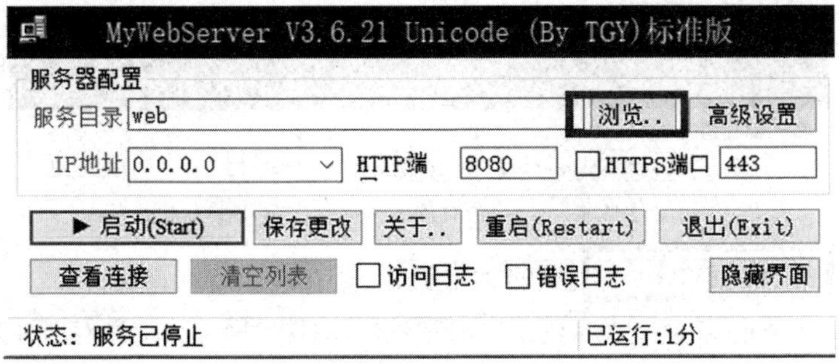

图 6-4

(5)选择测试网页文件,将服务目录定位于"测试主页"文件夹,如图 6-5
所示。

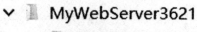

> ∨ 📁 **MyWebServer3621**
> 📄 **gziptmp**
> > 📁 **web**
> > 📁 **测试主页**

图 6-5

(6)点击"IP 地址"下拉菜单,选择测试地址(测试 PC 可能有多个地址,
请根据实际配置地址选择,下图仅为举例),如图 6-6 所示。

MyWebServer V3.6.21 Unicode (By TGY)标准版

服务器配置

服务目录 C:\Users\Mingwei\Desktop\MyWebServer3€ [浏览..] [高级设置]

IP地址 [10.10.243.9 ▼] HTTP端口 [8080] ☐ HTTPS端口 [443]

0.0.0.0
127.0.0.1
192.168.56.1
10.10.243.9

▶ 启 [关于..] [重启(Restart)] [退出(Exit)]

本系 ☐ 访问日志 ☐ 错误日志 [隐藏界面]

状态:服务已停止 已运行:0分

图 6-6

(7)点击启动,如图 6-7、图 6-8 所示。

MyWebServer V3.6.21 Unicode (By TGY)标准版

服务器配置

服务目录 C:\Users\Mingwei\Desktop\MyWebServer3€ [浏览..] [高级设置]

IP地址 [10.10.243.9 ▼] HTTP端 [8080] ☐ HTTPS端口 [443]

[▶ 启动(Start)] [保存更改] [关于..] [重启(Restart)] [退出(Exit)]

[查看连接] [清空列表] ☐ 访问日志 ☐ 错误日志 [隐藏界面]

状态:服务已停止 已运行:0分

图 6-7

图 6-8

（8）使用测试电脑访问该主机地址即端口号，即 X. X. X. X：8080 即可实现网页访问。

6.5　FTP 服务器搭建方法

本书实验涉及的 FTP 服务器中采用 FileZilla Server 软件进行搭建，具体方法如下：

（1）双击运行"FileZilla Server. exe"（Win Vista 及以上系统请使用"以管理员身份运行"，选择"是"安装服务及启动服务，如图 6-9 至图 6-13 所示。

图 6-9

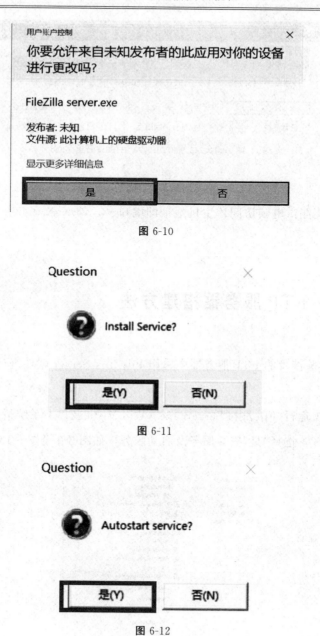

图 6-10

图 6-11

图 6-12

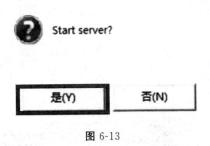

图 6-13

（2）双击运行"FileZilla Server Interface. exe"，会提示你连接到服务器，什么都不用设置，直接点"确定"进入运行界面，会提示你已成功连接服务器，如图 6-14、6-15 所示。

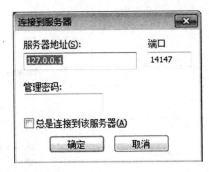

图 6-14

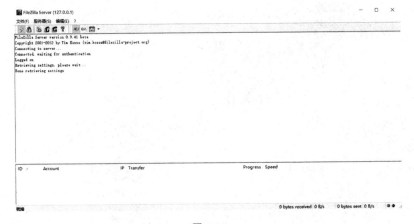

图 6-15

（3）点击"编辑"—"用户"，输入访问密码，添加一个用户，然后在"共享文件夹"下设置将要设为 FTP 目录的文件夹和操作权限，点击确定，如图 6-16 至图 6-18 所示。

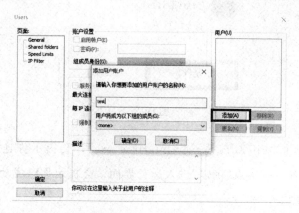

图 6-16

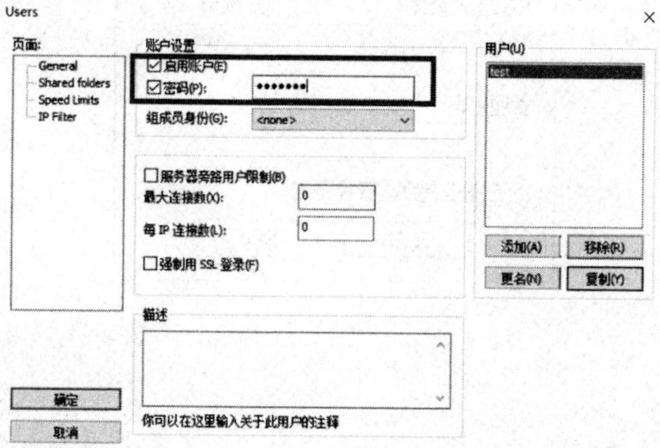

图 6-17

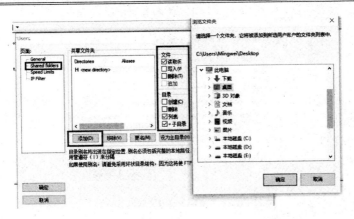

图 6-18

其他安全或功能方面的设置，请自行使用工具栏根据需要自行调整。

(4)使用测试 PC，通过[Windows＋R]打开运行窗口，输入 FTP：//×.×.×.×"，×.×.×.×"为已完成搭建的 FTP 服务器的 IP 地址，按回车就弹出了验证窗口。或者打开一个文件夹在地址栏直接输入 FTP：//×.×.×.×"并回车。然后输入刚才设置的用户和密码，回车，即可访问该 FTP 服务器(请保证 FTP 服务器上系统自带的防火墙关闭，否则无法正常访问)，如图 6-19、图 6-20 所示。

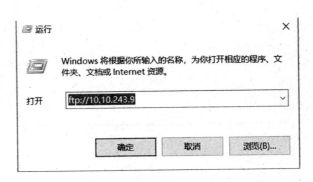

图 6-19

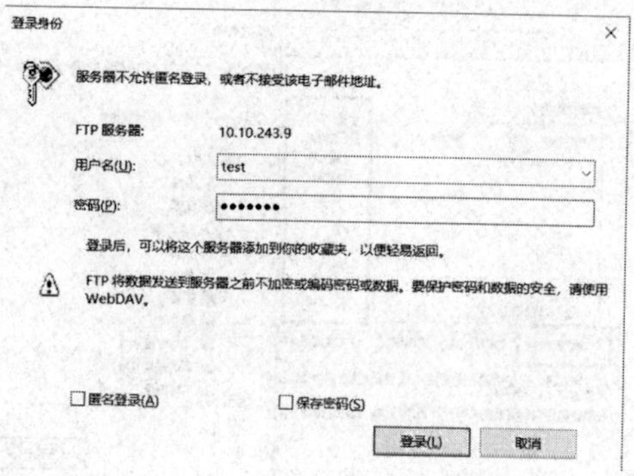

图 6-20

Lecture Notes
on Computer Network
Experiment

Wang Xueguang　著

吉林大学出版社

·长　春·

1　Introduction to laboratory environment

1.1　Laboratory layout，topological structure diagram and relevant explanation

Topology Map of Network and Information Security Laboratory of East China University of Political Science and Law（see Fig. 1-1）。

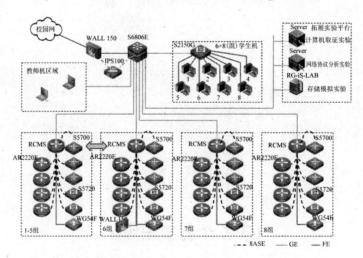

Fig. 1-1　Laboratory Topology Map

Each group of cabinets is consisted of four AR2220E routers，two S5720-

36C-EI-AC switches, and three S5700-28P-LI-AC. The diagram of specific cabinet layout is as shown in the Fig. 1-2.

前门

22	控制设备	22
21		21
20	理线架	20
19	理线架	19
18		18
17	AR2220E	17
16		16
15	AR2220E	15
14		14
13	AR2220E	13
12		12
11	AR2220E	11
10		10
9	S5720-36C-EI-AC	9
8		8
7	S5720-36C-EI-AC	7
6		6
5	S5700-28P-LI-AC	5
4		4
3	S5700-28P-LI-AC	3
2		2
1	S5700-28P-LI-AC	1

Fig. 1-2   Cabinet

1. 2   **Introduction to hardware structure**

1. 2. 1   AR2220E

Fig. 1-3 shows the AR2220E Router.

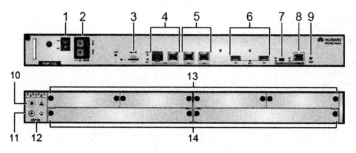

Fig. 1-3   AR2220E Router

Figure Note：

(1)Power switch

(2)DC power line interface

Explanation：

Connect the device to an external power supply using a DC power cable.

(3)Micro SD Card slot

(4)WAN side interface：GE Combo interface.

(5)WAN side interface：two GE electrical interfaces.

(6)Two USB interfaces（host）

Explanation：

When inserting a 3G USB modem，it is recommended to install a USB plasticcover（optional）to protect it. The 2 screw holes above the USB interface are used to fix the USB plastic cover. The appearance of a USB plastic cover is shown below：

(7)Mini USB interface

Explanation：

The Mini USB interface and the console interfacecan only be enabled for one at a time.

(8)CON/AUX interface

Explanation：

AR2220-DC does not support AUX function.

(9)RST button

Attention：

The Reset buttonis used for resetting the device manually. Resetting the device can cause a business interruption，so use the Reset button with caution.

(10)ESD jack

Explanation：

Wear an ESD wrist strap for the maintenance of the device. A section of the anti-static wristshould be inserted into the ESD jack.

(11)Grounding point

Explanation:

Use a grounding cableto reliably ground the device for lightning protection and anti-interference.

(12)Silk screen of the product model

(13)Four SIC slots

(14)Two WSIC slots

1. 2. 2   S-5700-28P-LI-AC

Fig. 1-4 shows the S-5700-28P-LI-AC Switch.

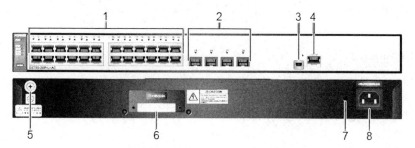

Fig. 1-4   S5700-28P-LI-AC Switch

Figure Note:

(1)Twenty-four 10/100/1000BASE-T Ethernet electrical interfaces.

(2)Four 1000BASE-X Ethernet optical interfaces.

Supported modules and cables:

GE optical module.

GE-CWDM colored light module.

GE-DWDM colored light module.

GE photoelectrical module (supported by V200R002C00 version and later versions, supporting 10M/100M/1000M rates).

Stack light module (supported by V200R007C00 version and later versions).

1m, 10m SFP+ high-speed cables.

3m, 10m AOC optical cables (supported by V200R003C00 version and later versions).

(3)One Mini USB interface

(4)One Console interface

(5)Grouping screw

Explanation:

Itis used with matching grounding cables.

(6)RPS power socket

Explanation:

Itis used with matching RPS cable，and the RPS cable does not support hot plugging.

(7)AC terminal anti-trip jack

Explanation:

To install the reserved AC terminal anti-trip jack，the AC terminal anti-tripis not shipped with the device.

(8)AC power socket

Explanation:

Itis used with matching AC power cables.

1.2.3　S5720-36C-EI-AC

Fig. 1-5 shows the S5720-36C-EI-AC Switch.

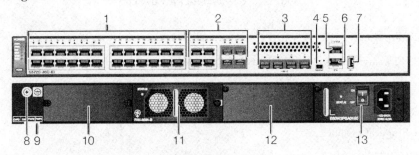

Fig. 1-5　S5720-36C-EI-AC Switch

Figure Note:

(1)Twenty-four 10/100/1000BASE-T Ethernet electrical interfaces.

(2)Four Combo interfaces（10/100/1000BASE-T+100/1000BASE-X）.

modules supported by Combo optimal interface:

FE optical module.

GE optical module.

GE-CWDM colored light module.

GE-DWDM colored light module.

(3)Four 10GE SFP+ Ethernet optical interfaces.

Supported modules and cables:

GE optical module.

GE-CWDM colored light module.

GE-DWDM colored light module.

GE photoelectrical module (only support gigabit rate).

10GE SPF+ optical module (not support OSXD22N00).

10GE-CWDM optical module.

10GE-DWDM optical module (supported by V200R009c00 version and later versions).

1m, 3m, 10m SFP+ high-speed cables.

5m SFP+ high-speed cables (supported by V200R009C00 version and later versions).

3m, 10m AOC optical cables.

(4)One Mini USB interface

(5)One console interface

Explanation:

Itis used with matching console cables. The consoling cables are not shipped with the device, if needed, please purchase it separately.

(6)One ETH management interface

(7)One USB interface

(8)Grouping screw

Explanation:

Itis used with matching grounding cables.

(9)Sequence number label

Explanation:

You can extract and view the serial number and MAC address information of the switch.

(10)Back card slot

Explanation:

Supported cards:

ES5D21X02S01 (two interfaces 10GE SFP+ optical interfacerear   card, used by S5720-EI series.)

ES5D21X02T01（two interfaces 10GBASE-T RJ45 electrical interface rear card，used by S5720-EI series.）

ES5D21VST000（two interfaces QSFP＋ dedicated stackrear　card，used by S5720-EI series）

(11)Fan module slot

Explanation：

Supported fan module：

FAN-028A-B fan module.

(12)Power module slot 2

Explanation：

Supported power module：

150W AC power module.

150W DC power module.

(13)Power module slot 1

Explanation：

Supported power module：

150W AC power module.

150W DC power module.

1.3　Instruction of experiment equipment diagram

Fig. 1-6 shows the diagram of the experiment equipment.

Fig. 1-6　Experimental Equipment Graph

1. 4   Instruction of each test-bed's login process

Process of PC's logging in the device through its Console interface.

1. 4. 1   Background information

After completing the configuration of Console user interface, you can login the device through Console interface. If the configuration device uses default properties of Console user interface, default AAA authentication method is used. The process of login is as follows.

1. 4. 2   Operating steps

(1) Plugging DB9 of Console communication cables into PC's port (COM) and plugging RJ-45 into Console interface as shown in the Fig. 1-7.

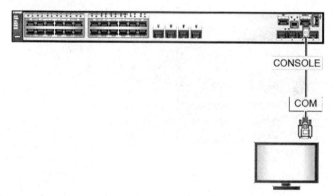

Fig. 1-7   Console Port Connection

(2) Launch the simulation software on PC, then establish a new connection and set connection interface and communication parameters. (The third party software SecureCRT is used as example for introduction here)

a. As shown in the following Fig. 1-8, click 🖳 , and establish a new connection.

Fig. 1-8　PC Terminal Simulation Software

b. As shown in the following Fig. 1-9, set the connection interface and communication parameters.

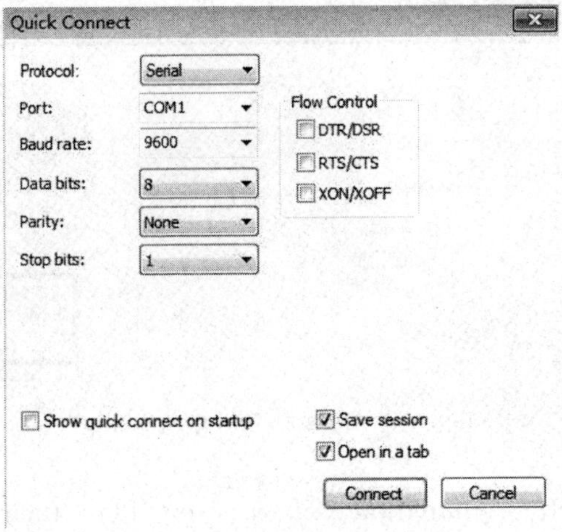

Fig. 1-9　Quick Connect

Please set the communication parameters of the terminal software consistent with the default value of the device: the transmission rate is 9600 bit/s , eight data bits, one stop bit, no parity and no flow control.

Instructions:

By default, the device does not have flow control. The RTS/CTS default is in the enabled state. Thus, this option should be checked, or the command line cannot be input in the terminal interface.

(3) Click "Connect" until the system appears the following display, reminding the user to input username and password. (When password authenticating, password is reminded to input. The following display information as an example.)

Login authentication

Username: admin

Password:

<HUAWEI>

(4)After accessing the device, the user can type commands to configure it. When help is needed, the user can type "?" at any time.

1.4.3 Check the result of configuration

Execute *display user* [*all*] command to check the user login information in the user interface.

Execute *display user-interface console* 0 command to check user interface information.

Execute *display local-user* command to check property information of local users.

Execute *display access-user* command to check user information of online connected users.

2 Basic equipment configuration

2.1 Use the command line management interface of the device

The device provides a wealth of functions, and accordingly provides a variety of configuration and query commands. To facilitate the use of these commands, Huawei switches register commands in different command line views according to functional classification. To configure a function, you must first enter the corresponding command line view, and then execute the corresponding command to configure it.

There are many command views provided by the device, and the one mentioned below is the most commonly used. How other views are entered is explained in specific commands.

2.1.1 Introduction to user interface

The system supports Console user interface and VTY user interface.

When a user logs in to a device using CLI, a user interface is assigned to manage and monitor the current session between the device and the user. Each user interface has a corresponding User-interface view, the User-interface view under the network administrator can configure a series of parameters, such as authentication model, user level, etc, when the User login using the User interface, it will be constrained by these parameters, so as to achieve the purpose of the unified management of User session connection.

The device supports two types of user interfaces:

Console user interface: it is used to manage and monitor users who log in through the Console port. The device provides a Console port of EIA/ TIA-232 DCE port type. The serial port of the user terminal can be directly connected to the device Console port to achieve local access to the device. MiniUSB also uses the Console interface to log in to the device.

Virtual Type Terminal user interface: it is used to manage and monitor users who log in via VTY. When a user establishes a Telnet or STelnet connection with a device through a terminal, a VTY channel is established. Currently, each device supports up to 15 VTY users to access at the same time.

2. 1. 1 The relationship between users and user interfaces

There is no fixed correspondence between user interface and user. User interface management and monitoring objects are users who log in in a certain way. Although a single user interface is used by only one user at a time, it is not specific to a user.

When the user logs in, the system will automatically assign the user an idle user interface of a certain type with the smallest number according to the user's login mode, and the whole login process will be constrained by the configuration under the user interface view. For example, when user A logs in to the device using the Console port, it will be constrained by the configuration under the Console user interface view, and when it logs in to the device using VTY 1, it will be constrained by the configuration under the VTY 1 user interface view. The same user login mode is different, the user interface assigned is different; The user interface may be assigned differently depending on the time the same user logs in.

📖 Instructions：

If a VTY user interface fails to respond to the device for a long time for two times, the VTY user interface will be locked, users can log in through other VTY user interfaces, and the device can be recovered after restart.

2.1.3 **Relative number**

The form of relative numbering is：User interface type＋Serial number.

This numbering can only specify one or a set of user interfaces of a particular type. The rules for relative numbering are as follows：

The number of the Console user interface：CON 0。

The number of the VTY user interface：The first one is VTY 0，the second one is VTY 1，and so on.

Absolute number

With absolute numbering，you can uniquely specify a user interface or a set of user interfaces. Use the command display user-interface to see which user interfaces the devices currently support and their absolute Numbers.

For a device，there is only one Console port user interface，but there are 20 VTY user interfaces. The maximum number of user interfaces can be set in the system view using the user-interface maximum-vty command. The default value is 5. VTY 16—VTY 20 always exist in the system and are not controlled by the user-interface maximum-vty command.

By default，the Console and VTY user interfaces are numbered in absolute Numbers in the system，as shown in the Table 2-1.

Table 2-1 **Number of User Interface**

The user interface	Instructions	Absolute number	Relative number
Console user interface	It is used to manage and monitor users who log in through the Console port or MiniUSB port	0	0
VTY user interface	It is used to manage and monitor users who log in by Telnet or STelnet	34—48, 50—54	The first one is VTY 0, the second one is VTY 1, and so on. VTY 0—4 channels exist by default.
		Among them, 49 are reserved, and 50—54 are reserved Numbers for network management	The absolute number 34—48 corresponds to the relative number VTY 0—VTY 14
			The absolute number 50—54 corresponds to the relative number VTY 16—VTY 20
			Where, VTY 15 is reserved, and VTY 16—VTY 20 are reserved Numbers for network management
			VTY 16—VTY 20 can only be used if all VTY 0—VTY 14 are occupied and the user has configured AAA authentication

2.1.4 User authentication of the user interface

After configuring the user authentication mode of the user interface, the system authenticates the user's identity when the user logs into the device.

There are three ways to authenticate users: AAA authentication, Password authentication, and None authentication.

AAA authentication: user name and password are required to log in. The device verifies that the information entered by the user is correct according to the configured AAA user name and password. If it is correct, login is allowed, otherwise login is refused.

Password authentication: also known as Password authentication, login to enter the correct Password authentication. If the user enters the same password as the authentication password configured for the device, login is allowed, otherwise login is refused.

None authentication: also known as non-authentication, you do not need to enter any authentication information when logging in, and you can log in the device directly.

⚠ Pay attention to:

For better security, it is recommended not to use None authentication.

Regardless of the authentication mode, when the user fails to log in to the device, the system will start the delayed login mechanism. After the first login fails, the user may log in again after a delay of 5 seconds. For every subsequent login failure, the delay time increases by 5 seconds, that is, the second login failure is delayed by 10 seconds, and the third login failure is delayed by 15 seconds.

2.1.5 The user level of the user interface

The system supports hierarchical management of logins. The level at which a user can access a command is determined by the user level.

If the user is authenticated with Password or None, the command level accessible to the user who logs in to the device is determined by the user interface level at which the user logs in.

If AAA authentication is applied to the user, the command level accessible to the user logged into the device is determined by the level of the local user in the AAA configuration information.

2.1.6 Common command-line views

Table 2-2 Common Command Line Views

The name of the view	Enter the view	The view function
The user view	After successful login to the device from the terminal, the user enters the user view, which is displayed on the screen	In the user view, the user can complete the functions of viewing the running state and statistics
	<HUAWEI>	
The system view	Under the user view, enter the command system-view and press enter to enter the system view	In the system view, users can configure system parameters and enter other functional configuration views through this view
	<HUAWEI> system-view	
	Enter system view, return user view with Ctrl+Z	
	HUAWEI	

Continued

The name of the view	Enter the view	The view function
The interface view	Use the "interface" command and specify the interface type and interface number to access the corresponding interface view	A view that configures interface parameters is called an interface view. In this view, interface related physical properties, link layer characteristics, IP address and other important parameters can be configured
	HUAWEI interface gigabitethernet X/Y/Z	
	HUAWEI-GigabitEthernetX/Y/Z	
	X/Y/Z is the number of interfaces to be configured, corresponding to "stack ID/ subcard number/interface serial number" respectively	
	The GigabitEthernet interface in the example above is just a schematic	

When using the command line, users can use online help to get real-time help without memorizing a large number of complex commands.

To get online help by typing "?", the user can type "?" at any time during command line input. Command-line online help can be divided into full help and partial help.

2.1.7　Full help

When a user enters a command, you can use the command line to fully assist in getting prompts for all keywords and parameters. The following are some fully helpful examples for your reference:

Under any command view, we can get all the commands under the command view and their simple descriptions by typing "?". Examples are as follows:

<HUAWEI> ?

User view commands:

backup   Backup electronic elabel

cd   Change current directory

check   Check information

clear   Clear information

clock   Specify the system clock

compare   Compare function

...

Type a partial keyword for a command followed by a space-delimited "?",
if the location is a keyword, list all the keywords and their simple descriptions.
For example, the following:

<HUAWEI> system-view

[HUAWEI] user-interface vty 0 4

[HUAWEI-ui-vty0-4] authentication-mode ?

aaa   AAA authentication, and this authentication mode is recommended

none   Login without checking

password   Authentication through the password of a user terminal interface

[HUAWEI-ui-vty0-4] authentication-mode aaa ?

<cr>

[HUAWEI-ui-vty0-4] authentication-mode aaa

Among them, "aaa" and "password" are keywords, and "authentication"
and "authentication through the password of a user terminal interface" are
descriptions of keywords.

"<cr>" means that there is no keyword or parameter in this position.
Just type enter to execute.

Type a partial keyword for a command followed by a space-delimited "?", if the location is a keyword, list all the keywords and their simple descriptions. For example, the following:

```
<HUAWEI> system-view
[HUAWEI] ftp timeout ?
INTEGER<1-35791>  The value of FTP timeout, the default value is
30 minutes
[HUAWEI] ftp timeout 35 ?
<cr>
[HUAWEI] ftp timeout 35
```

Among them, "INTEGER<1-35791>" is the description of parameter value, "The value of FTP timeout, the default value is 30 minutes" is a simple description of the function of the parameter.

2.1.8 Partial help

When a user enters a command, if he or she remembers only the first character or characters of the command keyword, he or she can use the command line section to help get prompts for all keywords that begin with that string. Several examples of partial help are given below for your reference:

Type a string followed by "?", which lists all the keywords that begin with the string. Examples are as follows:

```
<HUAWEI> d?
debugging                          delete
dir                                display
<HUAWEI> d
```

Type a string followed by "?", which lists all the keywords that begin with the string. Examples are as follows:

```
<HUAWEI> display b?
bpdu                    bridge
buffer
```

Enter the first few letters of a keyword of the command, press < TAB > key, can show the complete keyword, the premise is that these letters can only mark the keyword, otherwise, continuously press < TAB > key, can appear different keywords, users can choose the required keywords.

2. 2 View the system and configuration information of the device

2. 2. 1 View CPU utilization statistics for the device

```
<HUAWEI> display cpu-usage
CPU Usage Stat. Cycle: 60 (Second)
CPU Usage            : 20% Max: 99%
CPU Usage Stat. Time : 2013-10-23  10: 04: 45
CPU utilization for five seconds: 5%: one minute: 5%: five minutes: 5%
Max CPU Usage Stat. Time : 2013-10-21 16: 14: 00.


TaskName   CPU   Runtime(CPU Tick High/Tick Low)   Task Explanation
VIDL       80%   0/e3a150c  0          DOPRA IDLE
OS         10%   0/ bfb044  0          Operation System
1AGAGT     6%    0/         0          1AGAGT
AAA        2%    0/         1d4a       AAA   Authen Account Authorize
```

ACL	1%	0/	13362	ACL Access Control List
ADPT	1%	0/	0	ADPT Adapter
AGNT	0%	0/	0	AGNTSNMP agent task
AGT6	0%	0/	0	AGT6SNMP AGT6 task
ALM	0%	0/	0	ALM Alarm Management
ALS	0%	0/	527a3e	ALS Loss of Signal
AM	0%	0/	232cf	AM Address Management
APP	0%	0/	0	APP
ARP	0%	0/	36582	ARP
ASFI	0%	0/	0	ASFI
ASFM	0%	0/	0	ASFM
BATT	0%	0/	0	BATT Main Task
BFD	0%	0/	100f36	BFD Bidirection Forwarding Detect
BFDA	0%	0/	0	BFDA BFD Adapter

2.2.2 View memory utilization information for the current device

```
<HUAWEI> display memory-usage
Memory utilization statistics at 2008-12-15 15：17：42+08：00
System Total Memory Is：394152720 bytes
Total Memory Used Is：130975664 bytes
Memory Using Percentage Is：33%
```

2. 2. 3 View component information for the device

```
<HUAWEI> display device
S5700-52P-LI-AC's Device status:
Slot Sub  Type        Online   Power    Register    Status  Role

0-S5700-52P-LI   Present   PowerOn  Registered   Normal   Master
```

2. 2. 4 View device version information

```
<HUAWEI> display version
Huawei Versatile Routing Platform Software
VRP (R) software, Version 5. 160 (S6720 V200R010C00SPC300)
Copyright (C) 2000-2016 HUAWEI TECH CO., LTD
HUAWEI S6720-54C-EI-48S-AC Routing Switch uptime is 0 week, 0 day, 5
hours, 8 minutes

ES5D2S50Q002 1(Master)    : uptime is 0 week, 0 day, 5 hours, 6 minutes
DDR      Memory Size      : 2048        M bytes
FLASH   Memory Size       : 446         M bytes
Pcb             Version   : VER. B
BootROM        Version    : 020a. 0001
BootLoad       Version    : 020a. 0001
CPLD     Version : 0108
Software   Version : VRP (R) Software, Version 5. 160 (V200R010C00SPC300)
CARD1 information
Pcb        Version : ES5D21Q04Q01 VER. A
CPLD     Version : 0105
```

PWR2 information

Pcb   Version : PWR VER. A

FAN1 information

Pcb   Version : NA

2.2.5　View the configuration parameters currently in effect for the device

display current-configuration

3   Experiments related to Switch

3. 1   Construction of VLAN

3. 1. 1   Networking requirements

As is shown in the Fig. 3-1, a lot of users are connected to the Switch of the enterprise and users who have the same business are connected to the network of the enterprise.

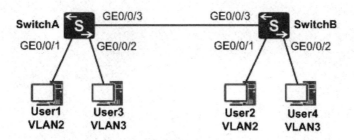

Fig. 3-1   Network Topology

The enterprise hope that users on the same business can have access to each other , while users on different business can't directly access to each other for the security of communication and avoidance of a broadcast storm.

VLAN partition based on interfaces is configured on the Switch. The interface which users on the same business are connected to can be partitioned to the same VLAN so that users belonging to different VLAN can't have direct access to two-layer communication and direct intercommunication can exist between users belonging to the same VLAN.

3.1.2 Configuration ideas

The ideas is adopted to configure VLAN as follows:

(1)Create a VLAN and add the user's interface to VLAN to achieve the two-layer flow isolation between users on different business.

(2)Configure the link type between SwitchA and SwitchB and VLAN passed so that users on the same business can communicate with each other through SwitchA and SwitchB.

3.1.3 Operating steps

(1)Create VLAN2 and VLAN3 on the SwitchA and add the interface connecting users to VLAN respectively. The configuration of SwitchB is similar with that of SwitchA. No more details.

```
<HUAWEI> system-view
[HUAWEI] sysname SwitchA
[SwitchA]vlan batch 2 3
[SwitchA]interface gigabitethernet 0/0/1
[SwitchA-GigabitEthernet0/0/1] port link-type access
[SwitchA-GigabitEthernet0/0/1] port default vlan 2
[SwitchA-GigabitEthernet0/0/1] quit
[SwitchA]interface gigabitethernet 0/0/2
[SwitchA-GigabitEthernet0/0/2] port link-type access
[SwitchA-GigabitEthernet0/0/2] port default vlan 3
```

[SwitchA-GigabitEthernet0/0/2] quit

(2) Configure the interface type connected to SwitchB on SwitchA and VLAN passed. The configuration of SwitchB is similar with that of SwitchA. No more details.

[SwitchA] interface gigabitethernet 0/0/3
[SwitchA-GigabitEthernet0/0/3] port link-type trunk
[SwitchA-GigabitEthernet0/0/3] port trunk allow-pass vlan 2 3

(3) Verify the result of configuration

Configure User1 and User2 on one network segment, such as 192. 168. 100. 0/24. Configure User3 and User4 on the other network segment, such as 192. 168. 200. 0/24.

User1 and User2 can Ping with each other, but are unable to Ping User3 and User4. User3 and User4 can Ping with each other, but are unable to Ping User1 and User2.

3. 1. 4　Configuration files

(1) The configuration file of SwitchA

```
#
sysname SwitchA
#
vlan batch 2 to 3
#
interface GigabitEthernet0/0/1
port link-type access
port default vlan 2
#
```

```
interface GigabitEthernet0/0/2
port link-type access
port default vlan 3
#
interface GigabitEthernet0/0/3
port link-type trunk
port trunk allow-pass vlan 2 to 3
#
return
```

(2)The configuration file of SwitchB

```
#
sysname SwitchB
#
vlan batch 2 to 3
#
interface GigabitEthernet0/0/1
port link-type access
port default vlan 2
#
interface GigabitEthernet0/0/2
port link-type access
port default vlan 3
#
interface GigabitEthernet0/0/3
port link-type trunk
port trunk allow-pass vlan 2 to 3
#
return
```

3. 2   **Isolation of Switch ports**

3. 2. 1   **Networking requirements**

The R&D(research and development) office workers in one company are divided into workers of their company, workers of cooperation partner A and workers of cooperation partner B. As is shown in the Fig. 3-2, PC1 and PC2 respectively represent partner A and partner B , and PC3 represents the company's R&D staff. The company hopes that under the premise of saving resources of VLAN, the company's staff can communicate with cooperation partner A and cooperation partner B.

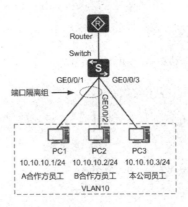

Fig. 3-2   Network Topology

3. 2. 2   **Configuration ideas**

The idea is adopted to configure ports isolation as follows

(1)Add an interface to VLAN.

(2)In the default mode of the device, port isolation is in the status of isolation in layer 2 and intercommunication in layer 3. The isolation of two-layer data within the isolation group can be achieved as long as the interface is added to the isolation group.

3.2.3 **Operating steps**

(1)Configure the function of ports isolation

Configure the function of ports isolation on GE0/0/1.

```
<HUAWEI> system-view
[HUAWEI] sysname Switch
[Switch]vlan 10
[Switch-vlan10]quit
[Switch]interface gigabitethernet 0/0/1
[Switch-GigabitEthernet0/0/1] port link-type access
[Switch-GigabitEthernet0/0/1] port default vlan 10
[Switch-GigabitEthernet0/0/1] port-isolate enable group 3
[Switch-GigabitEthernet0/0/1] quit
```

Configure the function of ports isolation on GE0/0/2.

```
[Switch]interface gigabitethernet 0/0/2
[Switch-GigabitEthernet0/0/2] port link-type access
[Switch-GigabitEthernet0/0/2] port default vlan 10
[Switch-GigabitEthernet0/0/2] port-isolate enable group 3
[Switch-GigabitEthernet0/0/2] quit
```

Add GE0/0/3 into VLAN 10.

```
[Switch]interface gigabitethernet 0/0/3
[Switch-GigabitEthernet0/0/3] port link-type access
[Switch-GigabitEthernet0/0/3] port default vlan 10
[Switch-GigabitEthernet0/0/3] quit
```

(2)Verify the result of configuration

Data packets of PC1 and PC2 can't be interworked.

Data packets of PC1 and PC3 can be interworked.

Data packets of PC2 and PC3 can be interworked.

3.2.4 **Configuration files**

The configuration file of Switch is as follows

```
#
sysname Switch
#
vlan batch 10
#
interface GigabitEthernet0/0/1
port link-type access
port default vlan 10
port-isolate enable group 3
#
interface GigabitEthernet0/0/2
port link-type access
port default vlan 10
port-isolate enable group 3
#
interface GigabitEthernet0/0/3
port link-type access
port default vlan 10
#
return
```

3.3 The port aggregation provides redundant backup links

3.3.1 Networking requirements

As is shown in Fig. 3-3，SwitchA and SwitchB are connected to the network of VLAN10 and VLAN20 respectively through the Ethernet link and there is large data traffic between SwitchA and SwitchB.

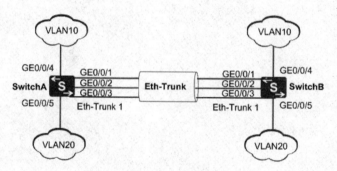

Fig. 3-3   Network Topology

Users hope that larger link bandwidth can be offered between SwitchA and SwitchB to allow users in the same VLAN to communicate with each other. At the same time，users hope that certain redundancy can be offered to ensure the reliability of data transmission and data link.

3.3.2 Configuration ideas

The idea of configuring load sharing link aggregation is adopted as follows：

(1)Create the interface of Eth-Trunk and add it to the member interface to increase the link bandwidth.

(2)Create VLAN and add the interface into VLAN.

(3)Configure load sharing mode to achieve load sharing of traffic between Eth-Trunk member interfaces and to increase reliability.

3. 3. 3 **Operating steps**

(1)Create the interface of Eth-Trunk on SwitchA and SwitchB and add it to member interfaces.

```
<HUAWEI> system-view
[HUAWEI] sysname SwitchA
[SwitchA] interface eth-trunk 1
[SwitchA-Eth-Trunk1] trunkport gigabitethernet 0/0/1 to 0/0/3
[SwitchA-Eth-Trunk1] quit
<HUAWEI> system-view
[HUAWEI] sysname SwitchB
[SwitchB] interface eth-trunk 1
[SwitchB-Eth-Trunk1] trunkport gigabitethernet 0/0/1 to 0/0/3
[SwitchB-Eth-Trunk1] quit
```

(2)Create VLAN and add the interface into VLAN.

\# CreateVLAN10 and VLAN20, then add them to member interfaces. The configuration of SwitchB is similar to that of SwitchA.

```
[SwitchA] vlan batch 10 20
[SwitchA] interface gigabitethernet 0/0/4
[SwitchA-GigabitEthernet0/0/4] port link-type trunk
[SwitchA-GigabitEthernet0/0/4] port trunk allow-pass vlan 10
[SwitchA-GigabitEthernet0/0/4] quit
[SwitchA] interface gigabitethernet 0/0/5
[SwitchA-GigabitEthernet0/0/5] port link-type trunk
[SwitchA-GigabitEthernet0/0/5] port trunk allow-pass vlan 20
[SwitchA-GigabitEthernet0/0/5] quit
```

\# Configure the Eth-Trunk1 interface to allow VLAN10 and VLAN20 to pass. The configuration of SwitchB is similar to that of SwitchA.

```
[SwitchA] interface eth-trunk 1
[SwitchA-Eth-Trunk1] port link-type trunk
[SwitchA-Eth-Trunk1] port trunk allow-pass vlan 10 20
[SwitchA-Eth-Trunk1] quit
```

(3)Configure the load sharing mode of Eth-Trunk 1. The configuration of SwitchB is similar to that of SwitchA. No more details.

```
[SwitchA] interface eth-trunk 1
[SwitchA-Eth-Trunk1] load-balance src-dst-mac
[SwitchA-Eth-Trunk1] quit
```

(4)Verify the result of configuration

Run "display eth-trunk 1" command in any view to check whether the Eth-Trunk is successfully created and the member interfaces are added correctly.

```
[SwitchA] display eth-trunk 1
Eth-Trunk1's state information is:
WorkingMode: NORMAL      Hash arithmetic: According to SA-XOR-DA
Least Active-linknumber: 1    Max Bandwidth-affected-linknumber: 8
Operate status: up           Number Of Up Port In Trunk: 3
```

PortName	Status	Weight
GigabitEthernet0/0/1	Up	1
GigabitEthernet0/0/2	Up	1
GigabitEthernet0/0/3	Up	1

GigabitEthernet 0/0/1, GigabitEthernet 0/0/2, and GigabitEthernet 0/

0/3. The "Operate status" of Eth-Trunk 1 is Up.

3.3.4　**Configuration files**

（1）Configuration file of SwitchA

```
#
sysname SwitchA
#
vlan batch 10 20
#
interface Eth-Trunk1
port link-type trunk
port trunk allow-pass vlan 10 20
load-balance src-dst-mac
#
interface GigabitEthernet0/0/1
eth-trunk 1
#
interface GigabitEthernet0/0/2
eth-trunk 1
#
interface GigabitEthernet0/0/3
eth-trunk 1
#
interface GigabitEthernet0/0/4
port link-type trunk
port trunk allow-pass vlan 10
#
interface GigabitEthernet0/0/5
```

```
port link-type trunk
port trunk allow-pass vlan 20
#
return
```

(2)**Configuration file of SwitchB**

```
#
sysname SwitchB
#
vlan batch 10 20
#
interface Eth-Trunk1
port link-type trunk
port trunk allow-pass vlan 10 20
load-balance src-dst-mac
#
interface GigabitEthernet0/0/1
eth-trunk 1
#
interface GigabitEthernet0/0/2
eth-trunk 1
#
interface GigabitEthernet0/0/3
eth-trunk 1
#
interface GigabitEthernet0/0/4
port link-type trunk
port trunk allow-pass vlan 10
```

```
#
interface GigabitEthernet0/0/5
port link-type trunk
port trunk allow-pass vlan 20
#
return
```

3. 4   Spanning tree configuration

3. 4. 1   Networking requirements

In a complex network, due to the need for redundant backup, network planners tend to deploy multiple physical links between devices, one for the primary link and the others for backup, which will inevitably result in the formation of loop network. If there is a loop in the network, it may cause a broadcast storm, and MAC items may be destroyed.

After planning the network, the network planner can deploy the STP (Spanning Tree Protocol) protocol to prevent loops in the network. When there is a loop in the network, STP blocks one of the ports to achieve the purpose of breaking the loop. As shown in the Fig. 3-4, there is a loop in the current network. SwitchA, SwitchB, SwitchC and SwitchD all run STP. By exchanging information with each other, they discover loop in the network and block a certain port selectively, finally prune the loop network into a loop-free tree network. It can prevent packets from circulating continuously in the loop network and devices won't decrease processing capacity due to repeated acceptance of the same message.

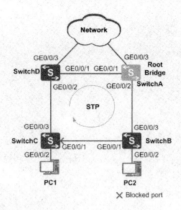

Fig. 3-4   Network Topology

3.4.2   **Configuration ideas**

The ideas are adopted to configure STP as follows:

Configure basic STP functions on the switching device in the ring network, including:

(1)Configure the devices STP in the loop network to work in STP mode.

(2)Configure root bridge and backing up the root bridge devices.

(3)Configure the path cost of the port to block the port.

(4)Enable STP to break the loop. The port connected to the PC does not participate in the STP calculation, and be set as an edge port Enable BPDU packet filtering on the port.

3.4.3   **Operating steps**

(1)Configure the basic function of STP

a. Configure the devices STP (Spanning Tree Protocol) in the loop network to work in STP mode.

Configure the STP working mode in SwitchA.

```
<HUAWEI>system-view
[HUAWEI]sysname SwitchA
[SwitchA]stp mode stp
```

Configure the STP working mode in SwitchB

```
<HUAWEI>system-view
[HUAWEI]sysname SwitchB
[SwitchB]stp mode stp
```

Configure the STP working mode in SwitchC

```
<HUAWEI>system-view
[HUAWEI]sysname SwitchC
[SwitchC]stp mode stp
```

Configure the STP working mode in SwitchD

```
<HUAWEI>system-view
[HUAWEI]sysname SwitchD
[SwitchD]stp mode stp
```

b. Configure root bridge and backing up the bridge root device.

Configure SwitchA as root bridge.

```
[SwitchA]stp root primary
```

Configure SwitchD as backup.

```
[SwitchD]stp root secondary
```

c. Configure the path cost of the port to block this port.

• The range of path cost is determined by the path cost calculation method. Here using Huawei calculation method as an example, we configure the path cost of the port being blocked to 20000.

• The port path cost of all switching devices in the same network should use the same calculation method.

\# Configure the path cost calculation method of SwitchA as Huawei calculation method.

```
[SwitchA]stp pathcost-standard legacy
```

\# Configure the path cost calculation method of SwitchB as Huawei calculation method.

```
[SwitchB]stp pathcost-standard legacy
```

\# Configure the path cost of the port GigabitEthernet0/0/1 in SwitchC to 20000.

```
[SwitchC]stp pathcost-standard legacy
[SwitchC]interface gigabitethernet 0/0/1
[SwitchC-GigabitEthernet0/0/1]stp cost 20000
[SwitchC-GigabitEthernet0/0/1]quit
```

\# Configure the path cost calculation method of SwitchD as Huawei calculation method.

```
[SwitchD]stp pathcost-standard legacy
```

d. Enable STP to break the loop

• Configure the port connected to the PC as the edge port and Enable the BPDU packet filtering function on the port.

\# Configure port GigabitEthernet0/0/2 in SwitchB as the edge port and Enable the BPDU packet filtering function.

```
[SwitchB]interface gigabitethernet 0/0/2
[SwitchB-GigabitEthernet0/0/2]stp edged-port enable
```

```
[SwitchB-GigabitEthernet0/0/2]stp bpdu-filter enable
[SwitchB-GigabitEthernet0/0/2]quit
```

Configure port GigabitEthernet0/0/2 in SwitchC as the edge port and Enable the BPDU packet filtering function.

```
[SwitchC]interface gigabitethernet 0/0/2
[SwitchC-GigabitEthernet0/0/2]stp edged-port enable
[SwitchC-GigabitEthernet0/0/2]stp bpdu-filter enable
[SwitchC-GigabitEthernet0/0/2]quit
```

- Enable devices STP globally.
- # Enable SwitchA STP globally.

```
[SwitchA]stp enable
```

Enable SwitchB STP globally.

```
[SwitchB]stp enable
```

Enable SwitchC STP globally.

```
[SwitchC]stp enable
```

Enable SwitchD STP globally.

```
[SwitchD]stp enable
```

(2)Verify the result of configuration

With the above configuration, perform the following operations to verify the configuration when the network calculation is stable.

Run "display stp brief" command on SwitchA to check the port status and port protection type. The results are as follows:

[SwitchA]display stp brief

MSTID	Port	Role	STP State	Protection
0	GigabitEthernet0/0/1	DESI	FORWARDING	NONE
0	GigabitEthernet0/0/2	DESI	FORWARDING	NONE

After SwitchA is configured as the root bridge, GigabitEthernet 0/0/2 and GigabitEthernet 0/0/1 that are connected to SwitchB and SwitchD are elected as the designated ports in the spanning tree calculation.

♯ Run "display stp interface gigabitEthernet0/0/1 brief" command on Switch B. The status of GigabitEthernet0/0/1 is displayed. The result is as follows:

[SwitchB]display stp interface gigabitethernet 0/0/1 brief

MSTID	Port	Role	STP State	Protection
0	GigabitEthernet0/0/1	DESI	FORWARDING	NONE

Port GigabitEthernet0/0/1 is the designated port in the spanning tree election and is in the FORWARDING state.

♯ Run "display stp brief" command on SwitchC to check the port status. The results are as follows:

[SwitchC]display stp brief

MSTID	Port	Role	STP State	Protection
0	GigabitEthernet0/0/1	ALTE	DISCARDING	NONE
0	GigabitEthernet0/0/3	ROOT	FORWARDING	NONE

Port GigabitEthernet0/0/3 becomes the root port in the spanning tree election and is in the FORWARDING state.

Port GigabitEthernet0/0/1 becomes an alternate port in the spanning tree election and is in the DISCARDING state.

3. 4. 4   **Configuration files**

(1)Configuration file of SwtichA

```
#
sysname SwitchA
#
stp mode stp
stp instance 0 root primary
stp pathcost-standard legacy
#
return
```

(2)Configuration file of SwtichB

```
#
sysname SwitchB
#
stp mode stp
```

```
stp pathcost-standard legacy
#
interface GigabitEthernet0/0/2
stp bpdu-filter enable
stp edged-port enable
#
return
```

(3)Configuration file of SwtichC

```
#
sysname SwitchC
```

```
#
stp mode stp
stp pathcost-standard legacy
#
interface GigabitEthernet0/0/1
stp instance 0 cost 20000
#
interface GigabitEthernet0/0/2
stp bpdu-filter enable
stp edged-port enable
#
return
```

(4)Configuration file of SwtichD

```
#
sysname SwitchD
#
stp mode stp
stp instance 0 root secondary
stp pathcost-standard legacy
#
return
```

4　Experiments related to routing

4.1　Experiment of static routing

4.1.1　Networking requirements

As is shown in Fig. 4-1, hosts in different network segments are connected by some Routers. And the requirement is to come to the result that any two hosts in different network segments can communicate with each other without configuring dynamic routing protocols.

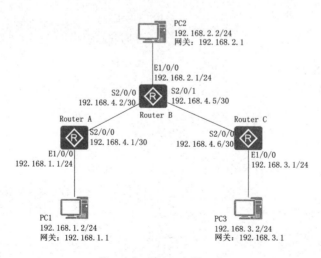

Fig. 4-1　Network Topology

4.1.2 **Configuration ideas**

Static routing is configured by the following ideas:

(1)Configure IP addresses for each router interface，so the networks are connected to each other.

(2)Configure the default gateway on each host IP address then configure IP static routing and default routing on each route so any two hosts in different network segments can communicate with each other without configuring dynamic routing protocols.

4.1.3 **Operating steps**

(1)Configure IP addresses for each router interface

♯ Configure Ip address for interface on Router A. Configuration of Router B and C is the same as Router A，so they are omitted.

[RouterA] interface Serial 2/0/0

[RouterA-Serial2/0/0] ip address 192. 168. 4. 1 30

[RouterA-Serial2/0/0] quit

[RouterA] interface ethernet 1/0/0

[RouterA-Ethernet1/0/0] ip address 192. 168. 1. 1 24

(2)Configure static Routing

♯ Configure IPv4 default routing on Router A.

[RouterA] ip route-static 0. 0. 0. 0 0. 0. 0. 0 192. 168. 4. 2

♯ Configure IPv4 default routing on RouterB.

[RouterB] ip route-static 192. 168. 1. 0 255. 255. 255. 0 192. 168. 4. 1

[RouterB] ip route-static 192. 168. 3. 0 255. 255. 255. 0 192. 168. 4. 6

♯ Configure IPv4 default routing on Router C

[RouterC] ip route-static 0. 0. 0. 0 0. 0. 0. 0 192. 168. 4. 5

(3)Configure the host

Configure default gateway of host PC1 as 192. 168. 1. 1, PC2 as 192.
168. 2. 1, PC3 as 192. 168. 3. 1

(4)Verify the configuration result

#Display the IP routing table of Router A.

[RouterA] display ip routing-table
Route Flags: R - relay, D - download to fib

Routing Tables: Public
      Destinations : 11   Routes : 11

Destination/Mask	Proto	Pre	Cos	Flags	NextHop	Interface
0. 0. 0. 0/0	Static	60	0	RD	192. 168. 4. 2	Serial2/0/0
192. 168. 1. 0/24	Direct	0	0	D	192. 168. 1. 1	Ethernet1/0/0
192. 168. 1. 1/32	Direct	0	0	D	127. 0. 0. 1	Ethernet1/0/0
192. 168. 1. 255/32	Direct	0	0	D	127. 0. 0. 1	Ethernet1/0/0
192. 168. 4. 1/30	Direct	0	0	D	192. 168. 4. 1	Serial2/0/0
192. 168. 4. 1/32	Direct	0	0	D	127. 0. 0. 1	Serial2/0/0
192. 168. 4. 255/32	Direct	0	0	D	127. 0. 0. 1	Serial2/0/0

127. 0. 0. 0/8	Direct	0	0	D	127. 0. 0. 1
InLoopBack0					
127. 0. 0. 1/32	Direct	0	0	D	127. 0. 0. 1
InLoopBack0					
127. 255. 255. 255/32	Direct	0	0	D	127. 0. 0. 1
InLoopBack0					
255. 255. 255. 255/32	Direct	0	0	D	127. 0. 0. 1
InLoopBack0					

Use the Ping command to verify connectivity.

[RouterA] ping 192. 168. 3. 1

PING 192. 168. 3. 1: 56 data bytes, press CTRL _ C to break

Reply from 192. 168. 3. 1: bytes=56 Sequence=1 ttl=254 time=62 ms

Reply from 192. 168. 3. 1: bytes=56 Sequence=2 ttl=254 time=63 ms

Reply from 192. 168. 3. 1: bytes=56 Sequence=3 ttl=254 time=63 ms

Reply from 192. 168. 3. 1: bytes=56 Sequence=4 ttl=254 time=62 ms

Reply from 192. 168. 3. 1: bytes=56 Sequence=5 ttl=254 time=62 ms

—192. 168. 3. 1 ping statistics—

5 packet(s) transmitted

5 packet(s) received

0. 00% packet loss

round-trip min/avg/max = 62/62/63 ms

Use the Tracert command to verify connectivity.

[RouterA] tracert 192. 168. 3. 1

traceroute to　192. 168. 3. 1(192. 168. 3. 1)，max hops: 30 ，packet length:

40，press CTRL _ C to break

1 192. 168. 4. 2 31 ms　32 ms　31 ms

2 192. 168. 4. 6 62 ms　63 ms　62 ms

4. 1. 4　Configuration files

(1)The configuration file of Router A

```
#
sysname RouterA
#
interface Ethernet1/0/0
ip address 192. 168. 1. 1 255. 255. 255. 0
#
interface Serial2/0/0
ip address 192. 168. 4. 1 255. 255. 255. 252
#
ip route-static 0. 0. 0. 0 0. 0. 0. 0 192. 168. 4. 2
#
return
```

(2)The configuration file of Router B

```
#
sysname RouterB
#
interface Serial2/0/0
ip address 192. 168. 4. 2 255. 255. 255. 252
```

```
#
interface Serial2/0/1
ip address 192.168.4.5 255.255.255.252
#
interface E1/0/0
ip address 192.168.2.1 255.255.255.0
#
ip route-static 192.168.1.0 255.255.255.0 192.168.4.1
ip route-static 192.168.3.0 255.255.255.0 192.168.4.6
#
return
```

(3)The configuration file of Router C

```
#
sysname RouterC
#
interface Ethernet1/0/0
ip address 192.168.3.1 255.255.255.0
#
interface Serial2/0/0
ip address 192.168.4.6 255.255.255.252
#
ip route-static 0.0.0.0 0.0.0.0 192.168.4.5
#return
```

4. 2   Experiment of RIP routing protocol

4. 2. 1   Networking requirements

As shown in the Fig. 4-2, there are four routers in the network, which are required to come to network interconnection on RouterA, RouterB, RouterC and RouterD.

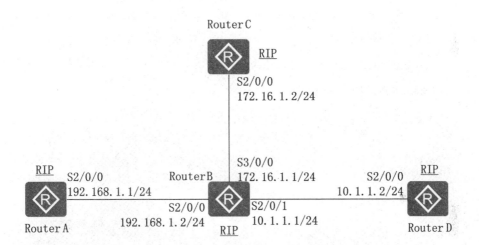

Fig. 4-2   Network Topology

4. 2. 2   Configuration ideas

RIP-2 routing protocol is recommended on the purpose of network interconnection of devices in small networks.

（1）Configure IP addresses of each interface to make the network accessible.

（2）Enable RIP on each router, and basically achieve network interconnection.

（3）Configure RIP-2 on all routers to improve the routing scalability of RIP.

4.2.3 **Operating steps**

（1）Configure IP addresses for each router interface

＃Configure RouterA.

```
[RouterA] interface Serial 2/0/0
[RouterA-Serial2/0/0] ip address 192.168.1.1 24
```

RouterB，RouterC and RouterD are configured in the same way as RouterA（omitted）.

（2）Configure basic functionality of RIP

＃Configure RouterA.

```
[RouterA] rip
[RouterA-rip-1] network 192.168.1.0
[RouterA-rip-1] quit
```

＃Configure RouterB.

```
[RouterB] rip
[RouterB-rip-1] network 192.168.1.0
[RouterB-rip-1] network 172.16.0.0
[RouterB-rip-1] network 10.0.0.0
[RouterB-rip-1] quit
```

＃Configure RouterC.

```
[RouterC] rip
[RouterC-rip-1] network 172.16.0.0
[RouterC-rip-1] quit
```

＃Configure RouterD.

```
[RouterD] rip
[RouterD-rip-1] network 10. 0. 0. 0
[RouterD-rip-1] quit
```

View the RIP routing table of RouterA.

```
[RouterA] display rip 1 route
Route Flags: R - RIP
             A - Aging, S - Suppressed, G - Garbage-collect
```

Peer 192. 168. 1. 2   on Serial2/0/0

Destination/Mask	Nexthop	Cost	Tag	Flags	Sec
10. 0. 0. 0/8	192. 168. 1. 2	1	0	RA	14
172. 16. 0. 0/16	192. 168. 1. 2	1	0	RA	14

As is shown in the routing table, RIP-1 publishes routing information using a natural mask.

(3)Configure the version of RIP

Configure RIP-2 on RouterA.

```
[RouterA] rip
[RouterA-rip-1] version 2
[RouterA-rip-1] quit
```

Configure RIP-2 on RouterB.

```
[RouterB] rip
[RouterB-rip-1] version 2
[RouterB-rip-1] quit
```

Configure RIP-2 on RouterC.

［RouterC］rip

［RouterC-rip-1］version 2

［RouterC-rip-1］quit

＃ Configure RIP-2 on RouterD.

［RouterD］rip

［RouterD-rip-1］version 2

［RouterD-rip-1］quit

(4)Verify results of configuration

＃View the RIP routing table of RouterA.

［RouterA］display rip 1 route

Route Flags：R - RIP

              A - Aging，S - Suppressed，G - Garbage-collect

Peer 192. 168. 1. 2   onSerial2/0/0

Destination/Mask	Nexthop	Cost	Tag	Flags	Sec
10. 1. 1. 0/24	192. 168. 1. 2	1	0	RA	32
172. 16. 1. 0/24	192. 168. 1. 2	1	0	RA	32

As is shown in the routing table，the routes published by the RIP-2 have more accurate subnet mask information.

4. 2. 4 Configuration files

(1)The configuration file of RouterA.

```
#
sysname RouterA
#
interface Serial2/0/0
```

```
ip address 192. 168. 1. 1 255. 255. 255. 0
#
rip 1
version 2
network 192. 168. 1. 0
#
return
```

(2)The configuration file of RouterB

```
#
sysname RouterB
#
interface Serial2/0/0
ip address 192. 168. 1. 2 255. 255. 255. 0
#
interface Serial3/0/0
ip address 172. 16. 1. 1 255. 255. 255. 0
#
interface Serial2/0/1
ip address 10. 1. 1. 1 255. 255. 255. 0
#
rip 1
version 2
network 192. 168. 1. 0
network 172. 16. 0. 0
network 10. 0. 0. 0
#
return
```

（3）The configuration file of RouterC

```
#
sysname RouterC
#
interface Serial2/0/0
ip address 172. 16. 1. 2 255. 255. 255. 0
#
rip 1
version 2
network 172. 16. 0. 0
#
return
```

（4）The configuration file of RouterD

```
#
sysname RouterD
#
interface Serial2/0/1
ip address 10. 1. 1. 2 255. 255. 255. 0
#
rip 1
version 2
network 10. 0. 0. 0
#
return
```

4.3   Experiment of OSPF routing protocol

4.3.1 Networking requirements

As shown in the Fig. 4-3, all routers run OSPF and divide the entire autonomous system into three regions, where RouterA and RouterB are used as ABR to forward routes between regions. Once configured, each router should learn the route to all network segments within the AS.

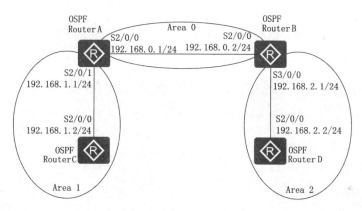

Fig. 4-3   Network Topology

4.3.2 Configuration ideas

The following ideas are adopted to configure the basic functions of OSPF:

(1)Enable OSPF on each router.

(2)Assign network segments in different regions.

4.3.3 Operating steps

(1)Configure IP addresses of each router interface

configure RouterA.

```
<Huawei>system-view
[Huawei]sysname RouterA
[RouterA] interface Serial 2/0/0
[RouterA-Serial2/0/0] ip address 192. 168. 0. 1 24
[RouterA-Serial2/0/0] quit
[RouterA] interface Serial 2/0/1
[RouterA-Serial2/0/1] ip address 192. 168. 1. 1 24
[RouterA-Serial2/0/1] quit
```

RouterB, RouterC and RouterD are configured in the same way as RouterA (omitted).

(2)Configure the basic functions of OSPF

#Configure RouterA.

```
[RouterA] router id 1. 1. 1. 1
[RouterA] ospf
[RouterA-ospf-1] area 0
[RouterA-ospf-1-area-0. 0. 0. 0] network 192. 168. 0. 0 0. 0. 0. 255
[RouterA-ospf-1-area-0. 0. 0. 0] quit
[RouterA-ospf-1] area 1
[RouterA-ospf-1-area-0. 0. 0. 1] network 192. 168. 1. 0 0. 0. 0. 255
[RouterA-ospf-1-area-0. 0. 0. 1] quit
[RouterA-ospf-1] quit
```

#Configure RouterB.

```
[RouterB] router id 2. 2. 2. 2
[RouterB] ospf
[RouterB-ospf-1] area 0
[RouterB-ospf-1-area-0. 0. 0. 0] network 192. 168. 0. 0 0. 0. 0. 255
[RouterB-ospf-1-area-0. 0. 0. 0] quit
[RouterB-ospf-1] area 2
```

```
[RouterB-ospf-1-area-0. 0. 0. 2] network 192. 168. 2. 0 0. 0. 0. 255

[RouterB-ospf-1-area-0. 0. 0. 2] quit

[RouterB-ospf-1] quit
```

Configure Router C.

```
[RouterC] router id 3. 3. 3. 3

[RouterC] ospf

[RouterC-ospf-1] area 1

[RouterC-ospf-1-area-0. 0. 0. 1] network 192. 168. 1. 0 0. 0. 0. 255

[RouterC-ospf-1-area-0. 0. 0. 1] quit

[RouterC-ospf-1] quit
```

Configure Router D.

```
[RouterD] router id 4. 4. 4. 4

[RouterD] ospf

[RouterD-ospf-1] area 2

[RouterD-ospf-1-area-0. 0. 0. 2] network 192. 168. 2. 0 0. 0. 0. 255

[RouterD-ospf-1-area-0. 0. 0. 2] quit

[RouterD-ospf-1] quit
```

(3) Verify results of configuration

View the OSPF neighbors of RouterA.

```
[RouterA] display ospf peer
                OSPF Process 1 with Router ID 1. 1. 1. 1
                            Neighbors
Area 0. 0. 0. 0 interface 192. 168. 0. 1(Serial2/0/0)'s neighbors
Router ID: 2. 2. 2. 2        Address: 192. 168. 0. 2
```

State: Full   Mode: Nbr is   Master   Priority: 1

DR: 192. 168. 0. 2   BDR: 192. 168. 0. 1   MTU: 0

Dead timer due in 36   sec

Retrans timer interval: 5

Neighbor is up for 00: 15: 04

Authentication Sequence: [0]

Neighbors

Area 0. 0. 0. 1 interface 192. 168. 1. 1(Serial2/0/1)'s neighbors

Router ID: 3. 3. 3. 3   Address: 192. 168. 1. 2

State: Full   Mode: Nbr is   Master   Priority: 1

DR: 192. 168. 1. 2   BDR: 192. 168. 1. 1   MTU: 0

Dead timer due in 39   sec

Retrans timer interval: 5

Neighbor is up for 00: 07: 32

Authentication Sequence: [0]

♯ Show the OSPF routing information of RouterA.

[RouterA] display ospf routing

OSPF Process 1 with Router ID 1. 1. 1. 1

Routing Tables

Routing for Network

Destination	Cost	Type	NextHop	AdvRouter	Area
192. 168. 0. 0/24	1	Transit	192. 168. 0. 1	1. 1. 1. 1	0. 0. 0. 0
192. 168. 1. 0/24	1	Transit	192. 168. 1. 1	1. 1. 1. 1	0. 0. 0. 1
192. 168. 2. 0/24	2	Inter-area	192. 168. 0. 2	2. 2. 2. 2	0. 0. 0. 0

Total Nets: 3

Intra Area: 2   Inter Area: 1   ASE: 0   NSSA: 0

Show the LSDB of RouterA.

```
[RouterA] display ospf lsdb
```

OSPF Process 1 with Router ID 1. 1. 1. 1

Link State Database

Area: 0. 0. 0. 0

Type	LinkState ID	AdvRouter	Age	Len	Sequence	Metric
Router	2. 2. 2. 2	2. 2. 2. 2	317	48	80000003	1
Router	1. 1. 1. 1	1. 1. 1. 1	316	48	80000002	1
Network	192. 168. 0. 2	2. 2. 2. 2	399	32	800000F8	0
Sum-Net	192. 168. 2. 0	2. 2. 2. 2	237	28	80000002	1
Sum-Net	192. 168. 1. 0	1. 1. 1. 1	295	28	80000002	1

Area: 0. 0. 0. 1

Type	LinkState ID	AdvRouter	Age	Len	Sequence	Metric
Router	3. 3. 3. 3	3. 3. 3. 3	217	60	80000008	1
Router	1. 1. 1. 1	1. 1. 1. 1	289	48	80000002	1
Network	192. 168. 1. 1	1. 1. 1. 1	202	28	80000002	0
Sum-Net	192. 168. 2. 0	1. 1. 1. 1	242	28	80000001	2
Sum-Net	192. 168. 0. 0	1. 1. 1. 1	300	28	80000001	1

View the routing table of RouterD, and use Ping to test connectivity.

```
[RouterD] display ospf routing
```

OSPF Process 1 with Router ID 4. 4. 4. 4

Routing Tables

Routing for Network

Destination	Cost	Type	NextHop	AdvRouter	Area
192. 168. 0. 0/242		Inter-area	192. 168. 2. 1	2. 2. 2. 2	0. 0. 0. 2
192. 168. 1. 0/243		Inter-area	192. 168. 2. 1	2. 2. 2. 2	0. 0. 0. 2
192. 168. 2. 0/241		Transit	192. 168. 2. 2	4. 4. 4. 4	0. 0. 0. 2

Total Nets: 5

Intra Area: 1   Inter Area: 2   ASE: 0   NSSA: 0

[RouterD] ping 192.168.1.2

PING 172.16.1.1: 56   data bytes, press CTRL _ C to break

Reply from192.168.1.2: bytes=56 Sequence=1 ttl=253 time=62 ms

Reply from192.168.1.2: bytes=56 Sequence=2 ttl=253 time=16 ms

Reply from192.168.1.2: bytes=56 Sequence=3 ttl=253 time=62 ms

Reply from192.168.1.2: bytes=56 Sequence=4 ttl=253 time=94 ms

Reply from192.168.1.2: bytes=56 Sequence=5 ttl=253 time=63 ms

——192.168.1.2 ping statistics ——

5 packet(s) transmitted

5 packet(s) received

0.00% packet loss

round-trip min/avg/max = 16/59/94 ms

4.3.4 Configuration files

(1)The configuration file of RouterA

\#

sysname RouterA

\#

router id 1.1.1.1

\#

interface Serial2/0/0

ip address 192.168.0.1 255.255.255.0

\#

interface Serial2/0/1

ip address 192.168.1.1 255.255.255.0

```
#
ospf 1
area 0. 0. 0. 0
network 192. 168. 0. 0 0. 0. 0. 255
area 0. 0. 0. 1
network 192. 168. 1. 0 0. 0. 0. 255
#
return
```

(2)The configuration file of RouterB

```
#
sysname Router B
#
router id 2. 2. 2. 2
#
interface Serial2/0/0
ip address 192. 168. 0. 2 255. 255. 255. 0
#
interface Serial2/0/1
ip address 192. 168. 2. 1 255. 255. 255. 0
#
ospf 1
area 0. 0. 0. 0
network 192. 168. 0. 0 0. 0. 0. 255
area 0. 0. 0. 2
network 192. 168. 2. 0 0. 0. 0. 255
#
return
```

(3)The configuration file of RouterC

```
#
sysname RouterC
#
router id 3. 3. 3. 3
#
interface Serial2/0/1
ip address 192. 168. 1. 2 255. 255. 255. 0
#
ospf 1
area 0. 0. 0. 1
network 192. 168. 1. 0 0. 0. 0. 255
#
return
```

(4)The configuration file of RouterD

```
#
sysname RouterD
#
router id 4. 4. 4. 4
#
interface Serial2/0/1
ip address 192. 168. 2. 2 255. 255. 255. 0
#
ospf 1
area 0. 0. 0. 2
network 192. 168. 2. 0 0. 0. 0. 255
#
return
```

4. 4   Communication and isolation between the same VLANs and different VLANs are achieved by three-layer switch

4. 4. 1   Networking requirements

As shown in the Fig. 4-4, for the security of communication, a company divides visitors, employees and servers into VLAN10, VLAN20 and VLAN30 respectively. The company wants that employees, server hosts, visitors can access the Internet. Visitors can only access the Internet and can't communicate with users from any other VLAN. Employee A can access to all resources in the server area, but other employees can only access port 21 of server A (FTP service).

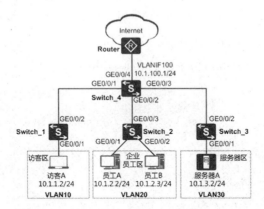

Fig. 4-4   Network Topology

4. 4. 2   Configuration ideas

The following ideas can be adopted to configure to achieve mutual visits control between VLAN through traffic policy:

(1)Configure VLAN and add each interface into VLAN, so as to make employees, servers and visitors on the layer-2 isolation.

(2) Configure VLANIF interface and its IP address, so as to make employees, servers and visitors on layer-3 connection.

(3)Configure uplink routing so as to employees, servers and visitors can access the Internet through Switch.

(4)Configure and apply the traffic policy to enable that employee A can access all resources in the server area. Other employees can only access port 21 of server A, and only employees are allowed to access the server; Enables only visitors to access the Internet.

4.4.3 **Operating steps**

(1) Configure VLAN and add each interface into VLAN, to make employees, servers and visitors on the layer-2 isolation.

♯ Create VLAN10 on Switch _ 1, and add interface GE0/0/1 into VLAN10 by the way of Untagged and add interface GE0/0/2 into VLAN10 by the way of Tagged. The configurations of Switch _ 2 and Switch _ 3 are similar to Switch _ 1, and shall not be described again.

```
<HUAWEI> system-view
[HUAWEI] sysname Switch _ 1
[Switch _ 1] vlan batch 10
[Switch _ 1] interface gigabitethernet 0/0/1
[Switch _ 1-GigabitEthernet0/0/1] port link-type access
[Switch _ 1-GigabitEthernet0/0/1] port default vlan 10
[Switch _ 1-GigabitEthernet0/0/1] quit
[Switch _ 1] interface gigabitethernet 0/0/2
[Switch _ 1-GigabitEthernet0/0/2] port link-type trunk
[Switch _ 1-GigabitEthernet0/0/2] port trunk allow-pass vlan 10
[Switch _ 1-GigabitEthernet0/0/2] quit
```

\# Create VLAN10, VLAN20, VLAN30 and VLAN100 on Switch _ 4, and configure the interface GE0/0/1—GE0/0/4 to add be Tagged with VLAN10, VLAN20, VLAN30, VLAN100.

```
<HUAWEI> system-view
[HUAWEI] sysname Switch _ 4
[Switch _ 4] vlan batch 10 20 30 100
[Switch _ 4] interface gigabitethernet 0/0/1
[Switch _ 4-GigabitEthernet0/0/1] port link-type trunk
[Switch _ 4-GigabitEthernet0/0/1] port trunk allow-pass vlan 10
[Switch _ 4-GigabitEthernet0/0/1] quit
[Switch _ 4] interface gigabitethernet 0/0/2
[Switch _ 4-GigabitEthernet0/0/2] port link-type trunk
[Switch _ 4-GigabitEthernet0/0/2] port trunk allow-pass vlan 20
[Switch _ 4-GigabitEthernet0/0/2] quit
[Switch _ 4] interface gigabitethernet 0/0/3
[Switch _ 4-GigabitEthernet0/0/3] port link-type trunk
[Switch _ 4-GigabitEthernet0/0/3] port trunk allow-pass vlan 30
[Switch _ 4-GigabitEthernet0/0/3] quit
[Switch _ 4] interface gigabitethernet 0/0/4
[Switch _ 4-GigabitEthernet0/0/4] port link-type trunk
[Switch _ 4-GigabitEthernet0/0/4] port trunk allow-pass vlan 100
[Switch _ 4-GigabitEthernet0/0/4] quit
```

(2) Configure VLANIF interface and its IP address, so as to make employees, servers and visitors on layer-3 connection.

\# Create VLANIF10, VLANIF20, VLANIF30 and VLANIF100 on Switch _ 4, and configure their IP addresses as 10. 1. 1. 1/24, 10. 1. 2. 1/24, 10. 1. 3. 1/24 and 10. 1. 100. 1/24 respectively.

```
[Switch _ 4] interface vlanif 10
[Switch _ 4-Vlanif10] ip address 10. 1. 1. 1 24
[Switch _ 4-Vlanif10] quit
[Switch _ 4] interface vlanif 20
[Switch _ 4-Vlanif20] ip address 10. 1. 2. 1 24
[Switch _ 4-Vlanif20] quit
[Switch _ 4] interface vlanif 30
[Switch _ 4-Vlanif30] ip address 10. 1. 3. 1 24
[Switch _ 4-Vlanif30] quit
[Switch _ 4] interface vlanif 100
[Switch _ 4-Vlanif100] ip address 10. 1. 100. 1 24
[Switch _ 4-Vlanif100] quit
```

(3)Configure uplink routing so as to employees, servers and visitors can access the Internet through Switch.

♯ Configure the basic functions of OSPF on Switch _ 4 , and publish the user network segment and the Internet segment between Switch _ 4 and Router.

```
[Switch _ 4] ospf
[Switch _ 4-ospf-1] area 0
[Switch _ 4-ospf-1-area-0. 0. 0. 0] network 10. 1. 1. 0 0. 0. 0. 255
[Switch _ 4-ospf-1-area-0. 0. 0. 0] network 10. 1. 2. 0 0. 0. 0. 255
[Switch _ 4-ospf-1-area-0. 0. 0. 0] network 10. 1. 3. 0 0. 0. 0. 255
[Switch _ 4-ospf-1-area-0. 0. 0. 0] network 10. 1. 100. 0 0. 0. 0. 255
[Switch _ 4-ospf-1-area-0. 0. 0. 0] quit
[Switch _ 4-ospf-1] quit
```

📖 Instructions:

The following configuration is required on Router:

• Add the interface of Switch to VLAN100 by the way of Tagged, and assign that the IP address of VLANIF100 is in the same segment as 10.1.100.1.

• Configure the basic functions of OSPF and publish the Internet segment between Switch and Router.

(4) Configure and apply the traffic policy to control access between employees, visitors, and servers

　a. Define each flow through ACL

　＃Configure ACL 3000 on Switch _ 4 to prevent visitors from accessing the employee zones and server zones.

[Switch _ 4] acl 3000

[Switch _ 4-acl-adv-3000] rule deny ip destination 10.1.2.1 0.0.0.255

[Switch _ 4-acl-adv-3000] rule deny ip destination 10.1.3.1 0.0.0.255

[Switch _ 4-acl-adv-3000] quit

　＃Configure ACL 3001 on Switch _ 4 so as to employee A can access all resources in the server area and other employees can only access port 21 of server A.

[Switch _ 4] acl 3001

[Switch _ 4-acl-adv-3001] rule permit tcp destination 10.1.3.2 0 destination-port eq 21

[Switch _ 4-acl-adv-3001] rule permit ip source 10.1.2.2 0 destination 10.1.3.1 0.0.0.255

[Switch _ 4-acl-adv-3001] rule deny ip destination 10.1.3.1 0.0.0.255

[Switch _ 4-acl-adv-3001] quit

　b. Configure flow classification to distinguish different flows

　＃Create c _ custom and c _ staff on Switch _ 4, and configure the matching rules 3000 and 3001 respectively.

```
[Switch _ 4] traffic classifier c _ custom
[Switch _ 4-classifier-c _ custom] if-match acl 3000
[Switch _ 4-classifier-c _ custom] quit
[Switch _ 4] traffic classifier c _ staff
[Switch _ 4-classifier-c _ staff] if-match acl 3001
[Switch _ 4-classifier-c _ staff] quit
```

c. Configure flow behavior , assign flow action

\# Create the b1 on Switch _ 4 and configure the allowed action.

```
[Switch _ 4] traffic behavior b1
[Switch _ 4-behavior-b1] permit
[Switch _ 4-behavior-b1] quit
```

d. Configure flow policy, association flow classification and

\# Create the flow policies p _ custom and p _ staff on Switch _ 4, and associate c _ custom and c _ staff with b1 respectively.

```
[Switch _ 4] traffic policy p _ custom
[Switch _ 4-trafficpolicy-p _ custom] classifier c _ custom behavior b1
[Switch _ 4-trafficpolicy-p _ custom] quit
[Switch _ 4] traffic policy p _ staff
[Switch _ 4-trafficpolicy-p _ staff] classifier c _ staff behavior b1
[Switch _ 4-trafficpolicy-p _ staff] quit
```

e. Apply flow policy to achieve access control among employees, visitors and servers

\# Apply the flow policy p _ custom and p _ staff in the incoming direction of VLAN10 and VLAN20 on Switch _ 4 respectively.

```
[Switch _ 4] vlan 10
[Switch _ 4-vlan10] traffic-policy p _ custom inbound
```

```
[Switch _ 4-vlan10] quit
[Switch _ 4] vlan 20
[Switch _ 4-vlan20] traffic-policy p _ staff inbound
[Switch _ 4-vlan20] quit
```

(5)Verify the results of configuration

Configure IP address of visitor A as 10. 1. 1. 2/24, and IP address of interface whose fault gateway is VLANIF10 is 10. 1. 1. 1;

Configure IP address of employee B as 10. 1. 2. 2/24, and IP address of interface whose fault gateway is VLANIF20 is 10. 1. 2. 1;

Configure IP address of employee C as 10. 1. 2. 3/24, and IP address of interface whose fault gateway is VLANIF20 is 10. 1. 2. 1;

Configure IP address of server A as 10. 1. 3. 2/24, and IP address of interface whose fault gateway is VLANIF30 is 10. 1. 3. 1;

After configuration finished:

• Visitor A cannot Ping employee A and server A; Employee A and server A cannot Ping visitor A.

• Employee A can Ping server A, that is, it can use server A and its FTP service.

• Employee B needn't Ping server A and can only use the FTP service of server A.

• Visitors, employee A, employee B and server A can all Ping the IP address 10. 1. 100. 2/24 of the interface that connects Router with Switch _ 4 , also can access the Internet.

4.4.4 **Configuration files**

(1) The configuration file of Switch_1

```
#
sysname Switch_1
#
vlan batch 10
#
interface GigabitEthernet0/0/1
port link-type access
port default vlan 10
#
interface GigabitEthernet0/0/2
port link-type trunk
port trunk allow-pass vlan 10
#
return
```

(2)The configuration file of Switch_2

```
#
sysname Switch_2
#
vlan batch 20
#
interface GigabitEthernet0/0/1
port link-type access
port default vlan 20
```

```
#
interface GigabitEthernet0/0/2
port link-type access
port default vlan 20
#
interface GigabitEthernet0/0/3
port link-type trunk
port trunk allow-pass vlan 20
#
return-
```

(3) The configuration file of Switch _ 3

```
#
sysname Switch _ 3
#
vlan batch 30
#
interface GigabitEthernet0/0/1
port link-type access
port default vlan 30 #
interface GigabitEthernet0/0/2
port link-type trunk
port trunk allow-pass vlan 30
#
return
```

(4) The configuration file of Switch _ 4

```
#
sysname Switch _ 4
#
vlan batch 10 20 30 100
#
acl number 3000
rule 5 deny ip destination 10. 1. 2. 0 0. 0. 0. 255
rule 10 deny ip destination 10. 1. 3. 0 0. 0. 0. 255
acl number 3001
rule 5 permit tcp destination 10. 1. 3. 2 0 destination-port eq ftp
rule 10 permit ip source 10. 1. 2. 2 0 destination 10. 1. 3. 0 0. 0. 0. 255
rule 15 deny ip destination 10. 1. 3. 0 0. 0. 0. 255
#
traffic classifier c _ custom operator and
if-match acl 3000
traffic classifier c _ staff operator and
if-match acl 3001
#
traffic behavior b1
permit
#
traffic policy p _ custom match-order config
classifier c _ custom behavior b1
traffic policy p _ staff match-order config
classifier c _ staff behavior b1
#
vlan 10
```

```
traffic-policy p _ custom inbound
vlan 20
traffic-policy p _ staff inbound
#
interface Vlanif10
ip address 10. 1. 1. 1 255. 255. 255. 0
#
interface Vlanif20
ip address 10. 1. 2. 1 255. 255. 255. 0
#
interface Vlanif30
ip address 10. 1. 3. 1 255. 255. 255. 0
#
interface Vlanif100
ip address 10. 1. 100. 1 255. 255. 255. 0
#
interface GigabitEthernet0/0/1
port link-type trunk
port trunk allow-pass vlan 10
#
interface GigabitEthernet0/0/2
port link-type trunk
port trunk allow-pass vlan 20
#
interface GigabitEthernet0/0/3
port link-type trunk
port trunk allow-pass vlan 30
#
```

```
interface GigabitEthernet0/0/4
port link-type trunk
port trunk allow-pass vlan 100
#
ospf 1
area 0.0.0.0
network 10.1.1.0 0.0.0.255
network 10.1.2.0 0.0.0.255
network 10.1.3.0 0.0.0.255
network 10.1.100.0 0.0.0.255
#
return
```

5　Network security ＋ Comprehensive experiment

5. 1　Using dynamic NAPT to realize LAN access to Internet

5. 1. 1　Networking requirements

As shown in the Fig. 5-1, private network users in Area A and Area B of a company are connected to the Internet. The public address of Serial 2/0/0 interface on router is 202. 169. 10. 1/2, while operator side address is 202. 169. 10. 2/24. Users in Area A want to use the address in the public network address pool (202. 169. 10. 100—202. 169. 10. 200) to replace the host address in Area A by NAT. (Section 192. 168. 20. 0/24) to address the Internet. Area B users want to combine the situation of less IP addresses in the public area network B. Use public network address pool (202. 169. 10. 80—202. 169. 10. 83) to replace the host address (10. 0. 0. 0/24) in Area B by replacing IP address and port to access the Internet.

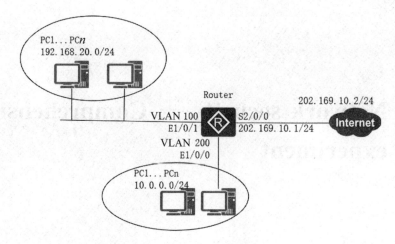

Fig. 5-1   Network Topology

5.1.2　Configuration ideas

The idea of configuring dynamic address translation is as follows:

Configure the IP address of interface, default routing and NAT Outbound under WAN side interface to realize the function of internal host accessing external network service.

5.1.3　Operating steps

(1) Configure interface IP address on Router

```
<Huawei> system-view
[Huawei] sysname Router
[Router]vlan 100
[Router-vlan100]quit
[Router]interface vlanif 100
[Router-Vlanif100]ip address 192.168.20.1 24
[Router-Vlanif100]quit
[Router]interface ethernet 1/0/1
[Router-Ethernet1/0/1]port link-type access
[Router-Ethernet1/0/1]port default vlan 100
```

```
[Router-Ethernet1/0/1]quit

[Router]vlan 200

[Router-vlan200]quit

[Router]interface vlanif 200

[Router-Vlanif200]ip address 10. 0. 0. 1 24

[Router-Vlanif200]quit

[Router]interface ethernet 1/0/0

[Router-Ethernet1/0/0]port link-type access

[Router-Ethernet1/0/0]port default vlan 200

[Router-Ethernet1/0/0]quit

[Router]interface Serial 2/0/0

[Router-Serial2/0/0]ip address 202. 169. 10. 1 24

[Router-Serial2/0/0] quit
```

(2) Configure default routing on Router, specifying the next hop address as 202. 169. 10. 2

```
[Router]ip route-static 0. 0. 0. 0 0. 0. 0. 0 202. 169. 10. 2
```

(3) Configure NAT Outbound on Router

```
[Router]nat address-group 1 202. 169. 10. 100 202. 169. 10. 200

[Router]nat address-group 2 202. 169. 10. 80 202. 169. 10. 83

[Router]acl 2000

[Router-acl-basic-2000]rule 5 permit source 192. 168. 20. 0 0. 0. 0. 255

[Router-acl-basic-2000]quit

[Router]acl 2001

[Router-acl-basic-2001]rule 5 permit source 10. 0. 0. 0 0. 0. 0. 255

[Router-acl-basic-2001]quit

[Router]interface Serial 2/0/0

[Router-Serial2/0/0]nat outbound 2000 address-group 1 no-pat
```

[Router-Serial2/0/0]nat outbound 2001 address-group 2

[Router-Serial2/0/0]quit

📖 Instructions：

If the ping -a source-ip-address command needs to be executed on the router to verify whether internal network users can access the Internet by sending the source IP address of the ICMP ECHO-REQUEST message，the ip soft-forward enhance enble command needs to be configured to enhance the forwarding function of the control message generated by the device，so that the source address of the private network can be converted into a public address through NAT. By default，the enhanced forwarding function of control messages generated by the device is enabled. If the undo ip soft-forward enhance enable command has already been executed to enhance the forwarding function，it is necessary to execute ip soft-forward enhance enable command again in the system view.

（4）Verify configuration results

＃ Execute the command "display NAT outbound" on Router to see the result of address translation.

<Router>display nat outbound

NAT Outbound Information：

Interface	Acl	Address-group/IP/Interface	Type
Serial2/0/0	2000	1	no-pat
Serial2/0/0	2001	2	pat

Total：2

Execute the command "ping" on Router to verify that the Intranet can access the Internet.

```
<Router>ping -a 192. 168. 20. 1 202. 169. 10. 2
PING 202. 169. 10. 2: 56 data bytes, press CTRL _ C to break
Reply from 202. 169. 10. 2: bytes=56 Sequence=1 ttl=255 time=1 ms
Reply from 202. 169. 10. 2: bytes=56 Sequence=2 ttl=255 time=1 ms
Reply from 202. 169. 10. 2: bytes=56 Sequence=3 ttl=255 time=1 ms
Reply from 202. 169. 10. 2: bytes=56 Sequence=4 ttl=255 time=1 ms
Reply from 202. 169. 10. 2: bytes=56 Sequence=5 ttl=255 time=1 ms
——202. 169. 10. 2 ping statistics——
5 packet(s) transmitted
5 packet(s) received
0. 00% packet loss
round-trip min/avg/max = 1/1/2 ms
<Router>ping -a 10. 0. 0. 1 202. 169. 10. 2
PING 202. 169. 10. 2: 56 data bytes, press CTRL _ C to break
Reply from 202. 169. 10. 2: bytes=56 Sequence=1 ttl=255 time=1 ms
Reply from 202. 169. 10. 2: bytes=56 Sequence=2 ttl=255 time=1 ms
Reply from 202. 169. 10. 2: bytes=56 Sequence=3 ttl=255 time=1 ms
Reply from 202. 169. 10. 2: bytes=56 Sequence=4 ttl=255 time=1 ms
Reply from 202. 169. 10. 2: bytes=56 Sequence=5 ttl=255 time=1 ms
——202. 169. 10. 2 ping statistics——
5 packet(s) transmitted
5 packet(s) received
0. 00% packet loss
round-trip min/avg/max = 1/1/2 ms
```

5.1.4 **Configuration files**

Router configuration file

```
#
sysname Router
#
vlan batch 100 200
#
acl number 2000
rule 5 permit source 192.168.20.0 0.0.0.255
#
acl number 2001
rule 5 permit source 10.0.0.0 0.0.0.255
#
nat address-group 1 202.169.10.100 202.169.10.200
nat address-group 2 202.169.10.80 202.169.10.83
#
interface Vlanif100
ip address 192.168.20.1 255.255.255.0
#
interface Vlanif200
ip address 10.0.0.1 255.255.255.0
#
interface Ethernet1/0/1
port link-type access
port default vlan 100
#
interface Ethernet1/0/0
```

```
port link-type access
port default vlan 200
#
interface Serial2/0/0
ip address 202. 169. 10. 1 255. 255. 255. 0
nat outbound 2000 address-group 1 no-pat
nat outbound 2001 address-group 2
#
ip route-static 0. 0. 0. 0 0. 0. 0. 0 202. 169. 10. 2
#
return
```

5. 2　Using NAT to achieve outer network access to Intranet server

5. 2. 1　Networking requirements

As shown in the Fig. 5-2, WWW Server and FTP Server are provided in a company's network for external network users to access, the internal IP address of WWW Server is 192. 168. 20. 2/24, the service port is 8080, and the published address is 202. 169. 10. 5/24. The internal IP address of FTP Server is 10. 0. 0. 3/24, the published address is 202. 169. 10. 33/24, and the opposite operator side address is 202. 169. 10. 2/24. The company's internal network is required to be connected to the Internet through the router's NAT function.

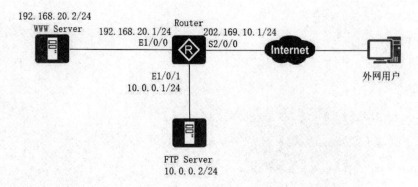

Fig. 5-2　Network Topology

5.2.2　**Configurationideas**

Using the following ideas to configure the internal server：

（1）Configure the IP address of the interface，and configure NAT Server on the Serial 2/0/0 interface to realize the function of external network users accessing the internal network server.

（2）Configure Router's default routing.

（3）NAT ALG function of FTP is enabled，and FTP access of external users can pass through NAT normally.

5.2.3　**Operating steps**

（1）Configure interface IP address and NAT Server on Router

```
<Huawei> system-view
[Huawei] sysname Router
[Router]vlan 100
[Router-vlan100]quit
[Router]interface vlanif 100
[Router-Vlanif100]ip address 192. 168. 20. 1 24
[Router-Vlanif100]quit
[Router]interface ethernet 1/0/0
[Router-Ethernet1/0/0]port link-type access
```

```
[Router-Ethernet1/0/0]port default vlan 100
[Router-Ethernet1/0/0]quit
[Router]vlan 200
[Router-vlan200]quit
[Router]interface vlanif 200
[Router-Vlanif200]ip address 10. 0. 0. 1 24
[Router-Vlanif200]quit
[Router]interface ethernet 1/0/1
[Router-Ethernet1/0/1]port link-type access
[Router-Ethernet1/0/1]port default vlan 200
[Router-Ethernet1/0/1]quit
[Router]interface Serial 2/0/0
[Router-Serial2/0/0]ip address 202. 169. 10. 1 24
[Router-Serial2/0/0]nat server protocol tcp global 202. 169. 10. 5 www inside
192. 168. 20. 2 8080
[Router-Serial2/0/0]nat server protocol tcp global 202. 169. 10. 33 ftp inside
10. 0. 0. 2 ftp
[Router-Serial2/0/0] quit
```

(2) Configure default routing on Router with the next hop address of 202. 169. 10. 2

```
[Router]ip route-static 0. 0. 0. 0 0. 0. 0. 0 202. 169. 10. 2
```

(3)NAT ALG function of FTP enabled on Router

```
[Router]nat alg ftp enable
```

(4)Verify configuration results

Execute the display NAT server operation on Router, and the results are as follows.

```
<Router>display nat server
Nat Server Information:
Interface: Serial 2/0/0
Global IP/Port              : 202.169.10.5/80(www)
Inside IP/Port             : 192.168.20.2/8080
Protocol : 6(tcp)
VPN instance-name          : ———————
Acl number                 : ———————
Vrrp id                    : ———————
Description                : ———————
Global IP/Port             : 202.169.10.33/21(ftp)
Inside IP/Port             : 10.0.0.2/21(ftp)
Protocol : 6(tcp)
VPN instance-name          : ———————
Acl number                 : ———————
Vrrp id                    : ———————
Description                : ———————
Total : 2
```

Execute the display NAT alg operation on Router, and the results are as follows.

```
<Router>display nat alg
NAT Application Level Gateway Information:
```

Application	Status
dns	Disabled
ftp	Enabled

rtsp	Disabled
sip	Disabled
pptp	Disabled

Verify that external network users can access the company's WWW Server and FTP Server properly (omitted).

5. 2. 4 **Configuration files**

Router configuration file

```
#
sysname Router
#
vlan batch 100 200
#
nat alg ftp enable
#
interface Vlanif100
ip address 192. 168. 20. 1 255. 255. 255. 0
#
interface Vlanif200
ip address 10. 0. 0. 1 255. 255. 255. 0
#
interface Ethernet1/0/0
port link-type access
port default vlan 100
#
interface Ethernet1/0/1
```

```
port link-type access
port default vlan 200
#
interface Serial 2/0/0
ip address 202. 169. 10. 1 255. 255. 255. 0
nat server protocol tcp global 202. 169. 10. 5 www inside 192. 168. 20. 2 8080
nat server protocol tcp global 202. 169. 10. 33 ftp inside 10. 0. 0. 2 ftp
#
ip route-static 0. 0. 0. 0 0. 0. 0. 0 202. 169. 10. 2
#
return
```

5.3　IP access list experiment

5.3.1　Networking requirements

As shown in the Fig. 5-3, an enterprise providing Web, FTP and Telnet services to the outside world accesses the external network through Router's interface Serial 2/0/0, and joins VLAN through Router's interface Eth1/0/0. The known enterprise segment is 202. 169. 10. 0/24. The IP addresses of WWW server, FTP server and Telnet server are 202. 169. 10. 5/24, 202. 169. 10. 6/24 and 202. 169. 10. 7/24 respectively.

In order to achieve high security of the internal network, enterprises want to configure the firewall function on Router so that only specific users can access the internal server in the external network and only internal servers can access the external network in the enterprise.

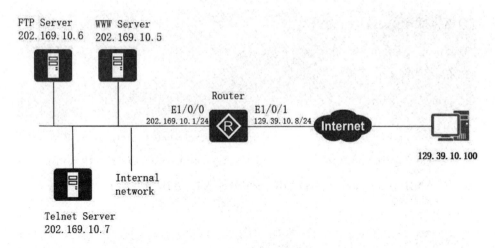

Fig. 5-3　Network Topology

5. 3. 1　**Configurationideas**

The following configuration ideas are adopted:

(1) Configure different security zones for internal and external networks of enterprises.

(2) Configure security domain and enable firewall function between security domains.

(3) Configure different advanced ACLs to categorize external network users who can access internal servers and internal servers who can access external networks.

(4) Configure packet filtering based on advanced ACL between security domains.

5. 3. 3　**Operating steps**

(1) Configure security area

Configure the security area for the internal network of the enterprise

```
<Huawei> system-view
[Huawei] sysname Router
[Router] firewall zone company
[Router-zone-company] priority 12
[Router-zone-company] quit
```

Configure the interface to add VLAN，and configure the IP address of the VLANIF interface，and add the interface VLANIF 100 to the security zone company.

```
[Router] vlan batch 100
[Router] interface ethernet 1/0/0
[Router-Ethernet1/0/0] port link-type access
[Router-Ethernet1/0/0] port default vlan 100
[Router-Ethernet1/0/0] quit
[Router] interface vlanif 100
[Router-Vlanif100] ip address 202.169.10.1 255.255.255.0
[Router-Vlanif100] zone company
[Router-Vlanif100] quit
```

Configure the security area for the external network.

```
[Router] firewall zone external
[Router-zone-external] priority 5
[Router-zone-external] quit
```

Add the interface Serial 2/0/0 to the security area external.

```
[Router] interface ethernet 1/0/1
[Router-Serial2/0/0] ip address 129.39.10.8 255.255.255.0
[Router-Serial2/0/0] zone external
[Router-Serial2/0/0] quit
```

(2)Configuring security domains

```
[Router] firewall interzone company external
[Router-interzone-company-external] firewall enable
[Router-interzone-company-external] quit
```

(3)Configure ACL 3001

Create ACL 3001.

```
[Router] acl 3001
```

Configuration allows specific users to access internal servers from external networks.

```
[Router-acl-adv-3001] rule permit tcp source 202. 39. 2. 3 0. 0. 0. 0 destination
202. 169. 10. 5 0. 0. 0. 0
[Router-acl-adv-3001] rule permit tcp source 202. 39. 2. 3 0. 0. 0. 0 destination
202. 169. 10. 6 0. 0. 0. 0
[Router-acl-adv-3001] rule permit tcp source 202. 39. 2. 3 0. 0. 0. 0 destination
202. 169. 10. 7 0. 0. 0. 0
```

Configure that other users cannot access any hosts within the enterprise from an external network.

```
[Router-acl-adv-3001] rule deny ip
[Router-acl-adv-3001] quit
```

(4)Configure ACL 3002

Create ACL 3002.

```
[Router] acl 3002
```

Configuration allows internal servers to access external networks.

```
[Router-acl-adv-3002] rule permit ip source 202. 169. 10. 5 0. 0. 0. 0
[Router-acl-adv-3002] rule permit ip source 202. 169. 10. 6 0. 0. 0. 0
[Router-acl-adv-3002] rule permit ip source 202. 169. 10. 7 0. 0. 0. 0
```

Configure that other users within the network cannot access the external network.

```
[Router-acl-adv-3002] rule deny ip
[Router-acl-adv-3002] quit
```

(5)Configure advanced ACL-based packet filtering between security domains

```
[Router] firewall interzone company external
[Router-interzone-company-external] packet-filter 3001 inbound
[Router-interzone-company-external] packet-filter 3002 outbound
[Router-interzone-company-external] quit
```

(6)Verify configuration results

After successful configuration，only specific hosts (129.39.10.100) can access internal servers and only internal servers can access external networks.

Execute the display firewall interzone [*zone-name* 1 *zone-name* 2] operation on Router，and the results are as follows.

```
[Router] display firewall interzone company external
interzone company external
firewall enable
packet-filter default deny inbound
packet-filter default permit outbound
packet-filter 3001 inbound
packet-filter 3002 outbound
```

5.3.4 Configuration files

Router configuration file

```
#
sysname Router
#
vlan batch 100
#
acl number 3001
rule 5 permit tcp source 202. 39. 2. 3 0 destination 202. 169. 10. 5 0
rule 10 permit tcp source 202. 39. 2. 3 0 destination 202. 169. 10. 6 0
rule 15 permit tcp source 202. 39. 2. 3 0 destination 202. 169. 10. 7 0
rule 20 deny ip
acl number 3002
rule 5 permit ip source 202. 169. 10. 5 0
rule 10 permit ip source 202. 169. 10. 6 0
rule 15 permit ip source 202. 169. 10. 7 0
rule 20 deny ip
#
interface Vlanif100
ip address 202. 169. 10. 1 255. 255. 255. 0
zone company
#
firewall zone company
priority 12
#
firewall zone external
priority 5
#
firewall interzone company external
firewall enable
```

```
packet-filter 3001 inbound
packet-filter 3002 outbound
#
interface Ethernet1/0/0
port link-type access
port default vlan 100
#
interface Ethernet1/0/1
ip address 129.39.10.8 255.255.255.0
zone external
#
return
```

6 Internal commands, using formats and server build introduction of Huawei switches and routers

6.1 Common commands

Table 6-1 Common Commands

CISCO	HUAWEI	Describe
no	undo	cancel, close current settings
show	display	display view
exit	quit	return to superior
hostname	sysname	setting host name
en, config terminal	system-view	enter global schema
delete	delete	deleted file
reload	reboot	restart
write	save	save the current configuration
username	local-user	create a user
shutdown	shutdown	disable, close port
show version	display version	display current system version
show startup-config	display saved-configuration	view saved configurations
show running-config	display current-configuration	display current configuration
no debug all	CTRL+D	cancel all DEBUG commands
erase startup-config	reset saved-configuration	delete configuration

Continued

CISCO	HUAWEI	Describe
end	return	back to the user view
exit	logout	logout
logging	info-center	specify information center configuration information
line	user-interface	enter line configuration (user interface) mode
start-config	saved-configuration	startup configuration
running-config	current-configuration	current configuration
host	IP host	the host name corresponds the IP address

6.2　Switching

Table 6-2　Switch Commands

CISCO	HUAWEI	Describe
enable password	set authentication password simple	configure plain text passwords
interface type/number	interface type/number	Access interface
interface VLAN 1	interface VLAN 1	enter the VLAN and configure VLAN management address
interface rang	interface ethID to ID	identify groups with multiple ports
enable secret	super password	setting a privileged password
duplex (half | full | auto)	duplex (half | full | auto)	configure interface state

Continued

CISCO	HUAWEI	Describe
speed (10/100/ 1000)	speed (10/100/1000)	configuration port rate
switchport mode trunk	port link-type trunk	configure Trunk
vlan ID /no vlan ID	vlan batch ID /undo vlan batch ID	add，delete VLAN
switchport access vlan	port acces vlan ID	access port to VLAN
show interface	display interface	view interface
show vlan ID	display vlan ID	view VLAN
encapsulation	link-protocol	encapsulation protocol
channel-group 1 mode on	port link-aggregation group 1	link aggregation
ip routing	default open	turn on the routing function of layer 3 switching
no switchport	non-support	opening interface three layer function
VTP domain	GVRP	dynamic registration and deletion of VLAN in Trans-Ethernet switch
spanning-tree VLAN ID root primary	STP instance ID root primary	STP configuration root bridge
spanning-tree VLAN ID priority	STP primary vlaue	configure bridge priority
show spanning-tree	dis STP brief	view the STP configuration

6.3　Routing

Table 6-3　Routing Commands

CISCO	HUAWEI	Describe
ip route 0. 0. 0. 0　　0. 0. 0. 0	ip　　route-static 0. 0. 0. 0 0. 0. 0. 0	configure default routing
ip route Target segment + mask off code　Next jump	ip　route-static　Target segment + mask off code Next jump	configure static routing
show ip route	display ip　routing-table	view the routing table
router rip /network network segment	rip　　　　　/network network segment	enable RIP，and declare a network segment
router ospf	ospf	enable OSPF
network ip radix-minus-one complement area ＜ area-id＞	area　＜area-id＞	configuring the OSPF region
no auto-summary	rip split-horizon	configuring RIP V2 horizontal segmentation
show ip protocol	display ip protocol	viewing routing protocols
access-list　1-99　permit/ deny IP	rule id permit source IP	standard access control list
access-list 100-199 permit/deny protocol source IP + radix-minus-one complement destination IP + radix-minus-one complement operator operan	rule｛ normal　｜　special｝ ｛permit　｜　deny｝｛tcp　｜ udp｝source｛＜ip wild＞｜ any｝destination ＜ ip wild ＞｜any｝［operate］	extended access control list
ip nat inside source　static	nat server global　＜ip＞ ［port］inside ＜ip＞ port ［protocol］	configure static address translation

6. 4 Method of building Web server

The Web server involved in this experiment is built with EasyServer software. The specific methods are as follows:

(1) Double-click to run the software program ▣ MyWebServer.exe ;

The following interface appears as shown in Fig. 6-1.

Fig. 6-1

(2) Click "Stop" as shown in Fig 6-2.

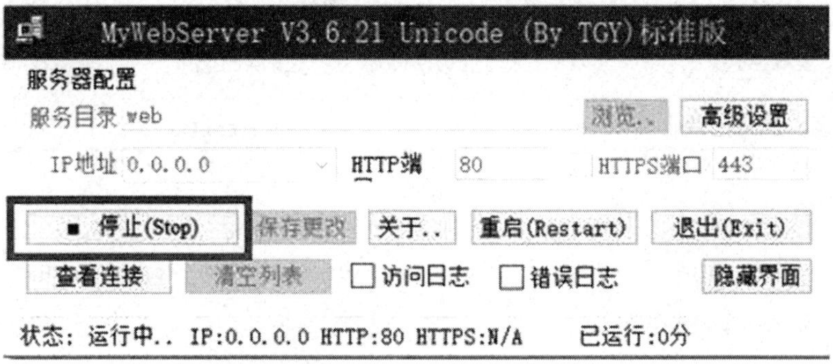

Fig. 6-2

（3）Based on the port number of the server in the experiment，modify the "HTTP port" parameter，such as port 8080 in the experiment，as shown in Fig. 6-3.

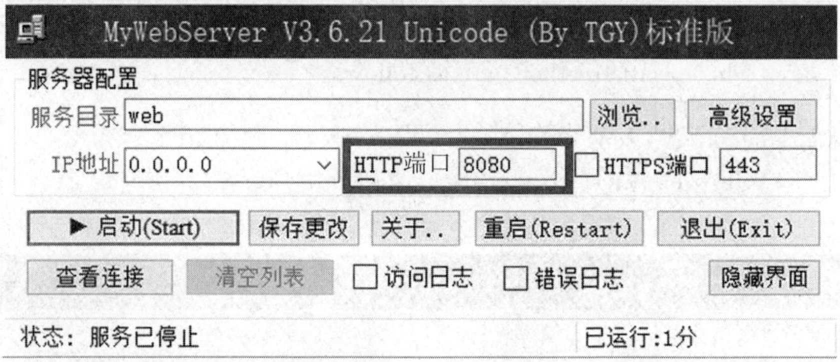

Fig. 6-3

（4）Modify the service directory and click the "Browse" option as shown in Fig. 6-4.

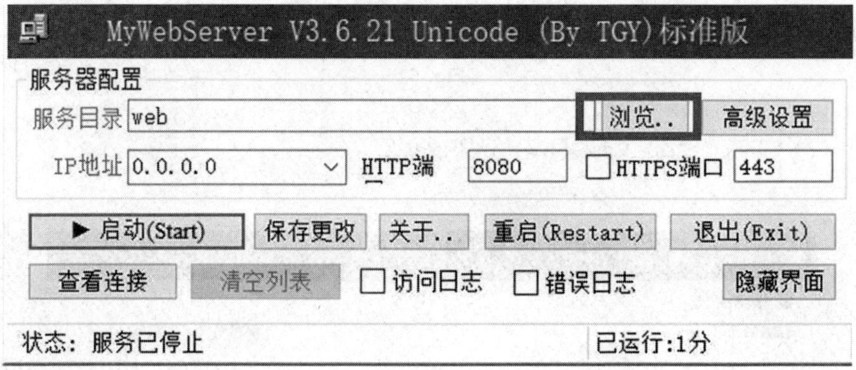

Fig. 6-4

（5）Select the test page file to locate the service directory in the "Test Home" folder as shown in Fig. 6-5.

Fig. 6-5

(6) Click the "IP address" drop-down menu to select the test address as shown in Fig. 6-6 (test PC may have multiple addresses，please select it according to the actual configuration address，below is only for example).

Fig. 6-6

(7)Click start as shown in Fig. 6-7 and Fig. 6-8.

Fig. 6-7

Fig. 6-8

(8)Use the test computer to access the host address，which is the port number like X. X. X. X：8080，to access to the Web page.

6. 5　Method of constructing FTP server

The FTP server involved in this experiment is built with FileZilla Server software. The specific methods are as follows：

1. Double-click to run "FileZilla Server. exe" (For Win Vista and above systems，use "run as an administrator"，and select "install service and start service"as shown from Fig. 6-9 to Fig. 6-13.

Fig. 6-9

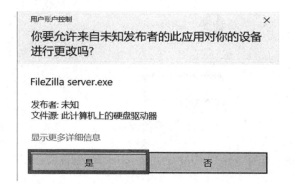

Fig. 6-10

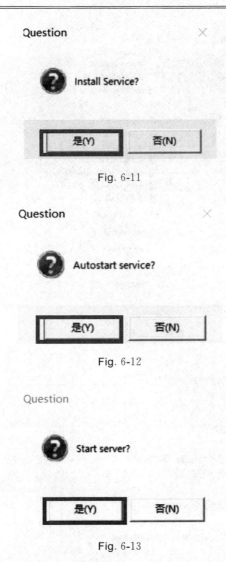

Fig. 6-11

Fig. 6-12

Fig. 6-13

(2) Double-click on "FileZilla Server Interface. exe", you will be prompted to connect to the server without setting up anything. Click "OK" directly to enter the running interface, and you will be prompted that you have successfully connected to the server as shown in Fig. 6-14 and Fig. 6-15.

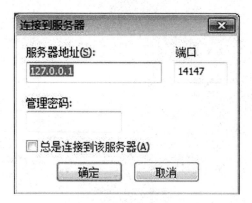

Fig. 6-14

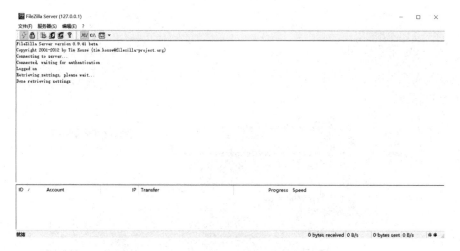

Fig. 6-15

(3) Click "Edit"-"User", enter the access password, add a user, and then, set the folder and operation permissions that will be set to the FTP directory under "shared folder". Click OK as shown from Fig. 6-16 to Fig. 6-18.

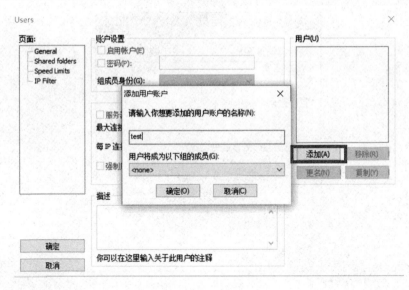

Fig. 6-16

Fig. 6-17

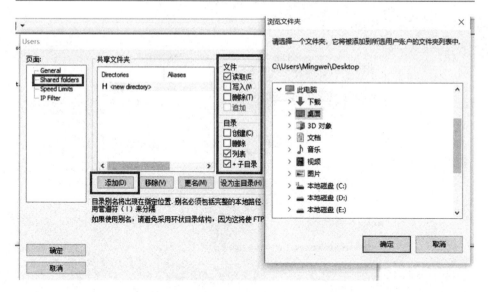

Fig. 6-18

For other security or functional settings，please use the toolbar and adjust it as needed.

（4）To use the test PC，open the run window through "Windows logo key ＋ R", type FTP：//×××（×××is the IP address of the completed FTP server)，and press "enter" to pop up the verification window. Or open a folder and type FTP：//××× directly in the address bar and press "enter". Then type the user and password set just now and press "enter"，you can access the FTP server as shown from Fig. 6-19 to Fig. 6-20（please make sure that the system firewall of the FTP server is shut down，otherwise you can not access it normally）

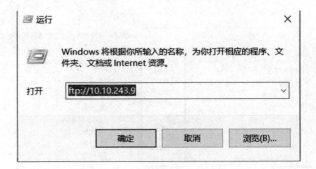

Fig. 6-19

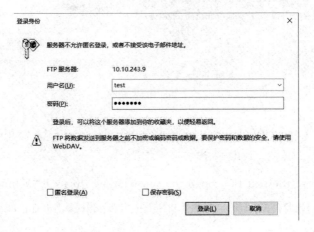

Fig. 6-20